St. Louis Community College

Forest Park
Florissant Valley
Meramec

Instructional Resources
St. Louis, Missouri

GAYLORD

SCIENCE AND SOCIETY

On Shifting Ground

THE STORY OF CONTINENTAL DRIFT

J. S. Kidd and Renee A. Kidd

☑ Facts On File, Inc.

This one is for Jack.

On Shifting Ground: The Story of Continental Drift

Copyright © 1997 by J. S. Kidd and Renee A. Kidd

All rights reserved. No part of this book may be reproduced or utilized in any form or by any means, electronic or mechanical, including photocopying, recording, or by any information storage or retrieval systems, without permission in writing from the publisher. For information contact:

Facts On File, Inc.
11 Penn Plaza
New York NY 10001

Library of Congress Cataloging-in-Publication Data

Kidd, J. S. (Jerry S.)
    On shifting ground : the story of continental drift / J. S. Kidd and Renee A. Kidd.
        p.   cm.—(Science and society)
    Includes bibliographical references (p.   –   ) and index.
    ISBN 0–8160–3582–2
    1. Continental drift. I. Kidd, Renee A. II. Title.
III. Series: Science and society (Facts on File, Inc.)
QE511.5.K475   1997
551.1
36–dc21   96–54299

Facts On File books are available at special discounts when purchased in bulk quantities for businesses, associations, institutions or sales promotions. Please call our Special Sales Department in New York at 212/967-8800 or 800/322-8755.

You can find Facts On File on the World Wide Web at http://www.factsonfile.com

Cover and text design by Cathy Rincon
Illustrations on pages 4, 6, 7, 9, 13, 15, 18, 20, 22, 35, 59, 71, 77, 86, 88, 95, 98, 110, and 113 by Marc Greene

This book is printed on acid-free paper.

Printed in the United States of America

MP FOF 10 9 8 7 6 5 4 3 2 1

Contents

PREFACE	v
ACKNOWLEDGMENTS	vii
INTRODUCTION	ix

1
WHAT IT IS — 1

2
THE SCIENCE BEGINS — 25

3
THE VISIONARY AND THE SKEPTICS — 31

4
THE NEW EARTH SCIENCES — 43

5
A TOUGH MAN FOR A TOUGH JOB — 49

6
SOME MAGNETIC PERSONALITIES — 63

7
MOVING ALONG TO LA JOLLA   79

8
THE GEOPOETS   90

9
A BIT OF GEOMETRIC SPICE   103

10
UPSTARTS AT LAMONT   106

11
THE ADVENTURE CONTINUES   115

    GLOSSARY   121

    FURTHER READING   128

    INDEX   129

Preface

This book is about the evolution of a set of ideas called continental drift theory. It reveals how modern scientific research led to major modifications of the original form of the theory. It describes how an intermediate version, called the theory of seafloor spreading, and the advanced version of these ideas called the theory of plate tectonics came to be accepted by the majority of earth scientists.

This book is also one in a series that deals with some of the connections between science and society. One goal of the series is to show how scientific facts are needed to help resolve many important public issues such as predicting natural disasters, improving the health of citizens, and protecting the environment.

Another goal is to show how society supports science. For example, the voters of the United States support scientific research as one means to achieve practical ends. However, the public also supports scientific research for the pure knowledge that is generated. Such support is mainly funneled through government agencies.

An example of government support of science is provided by arrangements begun during World War II. At that time, the U.S. Navy helped fund research in earth sciences that contributed to national security. This work also led directly to major revisions in the ideas about the formation of the earth and to a greater understanding of events such as earthquakes and volcanic eruptions.

In the mid-1960s, the results from many of these studies were brought together to yield the modern theory of plate tectonics. Applications of this theory have provided techniques to help locate off-shore oil fields. Thus, one broad program of scientific research in the earth sciences contributed to human knowledge in many different ways.

This book is designed to be used in schools and libraries to enrich the study of the earth sciences. Students can use it for supplementary readings and as a reference source. The book also serves as an historical narrative. A reader can learn about the backgrounds and experiences of the individuals who played key parts in the revision of the science of geology. This personalized approach might encourage some readers to enjoy the book simply as a good story.

Acknowledgments

We thank the administrators, faculty, and staff of two organizations devoted to higher education—The College of Library and Information Services of the University of Maryland, College Park and the Maryland College of Art and Design. In particular, Dean Ann Prentice and Associate Dean Diane Barlow at College Park have been extraordinarily patient and supportive.

We are also grateful for support and guidance from colleagues at the National Research Council/National Academy of Sciences in Washington, D.C. Again, special thanks to Anne Mavor and Alexandra Wigdor for their kindly dispositions and to Susan McCutchen for her high spirits.

As in our other projects, Sigrid Berge contributed her artistic skills of high merit.

Introduction

In June of 1991, two volcanic eruptions took place. The first event occurred at Mt. Unzen in Japan. The series of eruptions were very violent, and each explosion blew ash miles into the atmosphere. Toward the end of the eruptions, landslides of hot cinders killed more than 30 people. About 200 years before, when the population of Japan was less dense, a similar eruption of this volcano killed 15,000 people.

A week after the last eruptions of Mt. Unzen, Mt. Pinatubo on the island of Luzon in the Philippines exploded. Although Clark Air Force Base is nearby and the volcano's lower slope was heavily populated, only 50 people were killed in the disaster.

The low fatality rates were achieved because geologists used a combination of modern theories and newly developed instruments to forecast the times of the eruptions. If evacuations and restrictions on movement had been announced too early, serious economic losses and needless distress to the residents would have resulted. If the predictions had come too late, there would have been many more casualties. In both instances, accurate forecasts led to the timely evacuation of local people. Indeed, in the Philippines, the evacuation went smoothly because geologists were able to give an alert 48 hours before the eruptions. Twenty years ago, the ability to make such predictions was unknown.

Perhaps common sense should tell humans not to build towns near an active volcano. However, people do live near volcanoes and earthquake centers and have done so for thousands of years.

There are several reasons for such behavior. Climatic conditions near earthquake areas are often pleasant; volcanic neighborhoods usually have fertile soil; and such localities frequently have scenic views.

In one extreme case, the residents of Iceland heat their homes and obtain their electric power from boiling hot springs. These hot springs are produced by the same forces that make Iceland a center for both earthquakes and volcanoes. In 1783, 10,000 Icelanders died from a volcanic eruption, and over the intervening years, many other fatalities have occurred. However, the residents of this island believe that the benefits outweigh the risks.

For many years, scientists have sought to understand how such natural disasters occur. In one 20-year period, from 1946 to 1966, human knowledge of the earth increased very rapidly. At the end of the period, a new theory had replaced the traditional beliefs that had stood for two centuries. This theory allows a new understanding of earthquakes and volcanoes and opens the way for the use of new techniques in predicting possible disasters. The same set of ideas is now being applied to the search for oil and valuable minerals. Indeed, the ideas are also useful in solving other mysteries. For example, the theory explains how fossil remains of tropical plants and dinosaur bones came to be encased in the rocks of Antarctica.

Scientists did not develop their new ideas about the earth in a smooth series of advances. Progress took place in fits and starts from the beginning of the present century. In those earlier days, the rather radical new ideas were called the theory of continental drift. Theories are attempts to explain things. They must be proven before they can be considered rules or laws. The theory of continental drift was an attempt to explain the present location of the continents. It was put forward as a way specifically to explain the geographic relationship between the continents of Africa and South America—particularly the apparent fit between their Atlantic shorelines. In the late 1950s and early 1960s, similar but more advanced ideas were combined into the theory of seafloor spreading. After 1966, scientists used

Mt. St. Helens is in Washington State. It erupted in May, 1980. The source of the volcano's power is explained by the theory of plate tectonics. (Courtesy of the U.S. Geological Survey)

the accumulated evidence to promote the belief that the continents are in motion. This well-supported conviction is called the theory of plate tectonics.

During the early 1900s, most of the people who favored the new theory of continental drift received very little support from their scientific colleagues. Some advocates were virtually scorned. Others who agreed with the new ideas remained silent about their beliefs. Even after the 1970s, when reasonable evidence had been provided, some scientists were still very reluctant to give up their older notions.

By now, any modern book on the earth sciences will provide a discussion of plate tectonic theory and describe the facts upon which that theory is based. However, rarely will such a book provide information about the people who discovered the facts and recognized how the facts fit together. This book focuses on such people and their experiences in shaping a major scientific advance.

Science and Government

This story also reveals some of the ways in which the U.S. government encourages the conduct of scientific research. Substantial support for the earth sciences on the part of public officials began about 100 years ago with the need to find suitable sites for reservoirs and routes for highways and railroads. In the 1940s, attention shifted to the geology of the oceans—particularly, the structure and composition of the ocean floor. The official promotion of this kind of geological research was initiated by military leaders during World War II. Later, such research was linked to U.S. engagement in the cold war. The results of these studies were intended to provide a national security advantage to the United States and its allies by improving their ability to conduct submarine warfare. As the research began to generate promising new ideas, civilian agencies such as the National Aeronautics and Space Administration and the National Science Foundation also provided funding support.

The actions of both military and civilian agencies of the government were fostered by the U.S. Congress. The Congress has the responsibility for generating the nation's overall policies for science. These policies reflect the attitudes and opinions of the general public.

The large majority of citizens hold favorable attitudes toward science. The people's confidence in the earth sciences has been justified by advances in U.S. national security capabilities, improvements in the ability to respond to earthquakes and volcanic eruptions, and progress in understanding the earth.

1
What It Is

Most of us find our lives full of changes. Our day-to-day experiences are at least partly unpredictable. Many of us experience the ups and downs of daily life as a source of stress. We seek greater stability. We look for solid footing. We are more confident of our position when we have both feet on firm, stable ground. Indeed, the immobile Rock of Gibraltar is used to symbolize dependability.

Once you have actually experienced an earthquake or a volcanic eruption, the image of a stable earth loses some of its power. You begin to realize that the earth is not so reliable after all. Following such an event, it becomes easier to believe that the ground beneath us is in motion all the time. In fact, if you live in North America, you, your house, your family, and everything else is moving northwestward a bit more than an inch each year. You live on shifting ground.

The forces that move the whole continent every year come from the same energy sources that cause volcanic eruptions and earthquakes. Millions of years ago, these same forces created veins of metallic ore and trapped the raw materials that were transformed into petroleum and natural gas reserves. The forces, as it turns out, are not difficult to understand. They require just two basic conditions—heat and time—lots of heat and lots and lots of time.

An explanation of these forces and their consequences is provided by a set of ideas called the theory of plate tectonics. The world *tectonics* comes from the Greek word for "builder" and is used to imply action or change. In geology, the term suggests large changes in the earth's profile, such as the formation of mountain chains. The phrase *plate tectonics*, therefore, suggests that some sort of "plate" is active in some sort of geological development. What are these "plates"?

The Basic Concepts

People now know that the earth's crust, the rock that makes up the top layer of the planet, is divided into sections—or plates. The crust of the continents is composed mainly of granite, a comparatively lightweight rock. It can be more than 40 miles (64 km) in depth in some places.

Under the oceans, the crust is rather thin—no more than 5 or 6 miles (8 or 9.6 km) thick. The oceanic crust is composed of basaltic rock, which is heavier than the granite of the continental crust. Usually, this basalt is covered with sediment from the shells of dead sea creatures and other deposits. The bottom deposit of sediment is often compressed into hard limestone.

Although some plates are composed entirely of oceanic crust, most plates are composed of both kinds of crust. That is, most plates include some continental crust and a large area of oceanic crust.

Scientists have identified 14 of the plates that divide the earth's crust. Some are very large—such as the Pacific and Eurasian plates. Some are relatively small—like the Cocos and the Anatolian plates. The boundaries where the plates meet are almost always under the oceans and are geologically active. These plate boundaries generate many earthquakes. Fortunately, these earthquakes are usually small and occur several miles below the earth's surface.

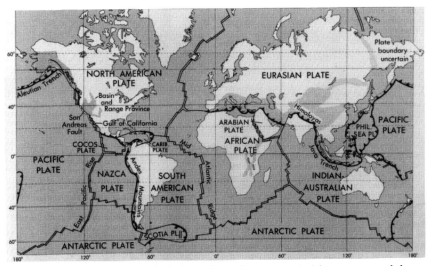

The boundaries of the crustal plates have been revealed by surveys of the ocean floors. Many earthquakes and volcanic eruptions occur along these edges. (Courtesy of the U.S. Geological Survey)

Scientists now know that all of these plates are on the move. However, different plates move at different speeds and in different directions. Even the maximum rate of movement is very slow—a few inches a year. Thus, in one person's lifetime, the Hawaiian Islands might move about 6 feet (1.8 m) to the northwest.

It is now virtually certain that North and South America were once joined together with Europe and Africa. The present separation of these continents has taken place over a time span of about 150 to 180 million years. When the separation began, the movement probably was faster than it is now.

To understand how this movement takes place, one must study the composition of the planet. Geologists know that the earth is made up of separate layers of rock. Below the crust—the uppermost layer—is the mantle, which is made up of three sections. The top section of the mantle has a thickness of 50 miles (80 km) or more. This layer of rock is quite hot, can be deformed like putty, and can be pushed into any open or uneven spaces. However, the rock is not a fluid. The asthenosphere lies below this top section of the mantle and is thought to be about 120 miles (192 km) thick. Here, rocks are far hotter and more fluid than the overlaying segment. The asthenosphere is a basic source of molten rock known as magma—the same material that streams from volcanoes as lava.

Below the asthenosphere, beginning at a depth of about 180 miles (288 km) and extending downward for hundreds

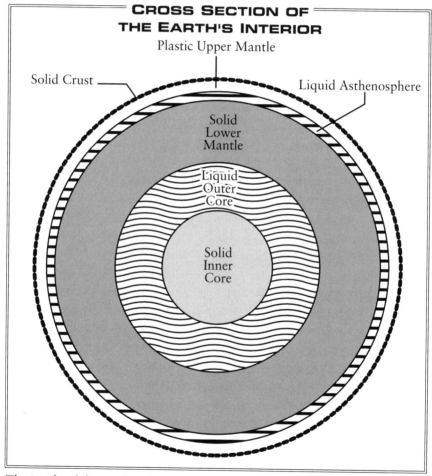

The inside of the earth is formed by several distinct layers. The central core is thought to be composed mainly of iron. Its rotational movement probably generates the earth's magnetic field.

of miles, is the third portion of the mantle. The high pressure at this depth, caused by the enormous weight of the rocks above, prevents the extremely high temperatures from melting the rock. Thus, the lowest part of the mantle is rather stiff.

Below the mantle is the core—the center of the earth. This structure, enveloped by the mantle, lies about 1,800 miles (2,880 km) below the crust and is a storehouse for the earth's heat. The core—which is not well understood—appears to be formed of two spherical sections. The outer section is probably fluid, and the inner section may be a solid ball of iron and other heavy metals. The round core apparently rotates inside the earth. Scientists are uncertain about the origin of this movement, but they believe that the rotation began when the earth solidified, about 4.5 billion years ago. This rotation causes the center of the earth to resemble an electric generator. The electric current thought to be produced by the spinning of the core is the source of the earth's magnetic field.

Plate Movement

The mantle above the core is lumpy. That is, parts of it are denser than others. This unevenness allows the heat of the core to rise along certain routes through the asthenosphere and upper mantle to the undersurface of the crust. These rising columns and curtains of heat, called plumes, drive magma from the asthenosphere up to and through the crust. The minor effects of this action can be seen on the earth's surface as hot springs and volcanoes.

Under the South Atlantic Ocean, between Africa and the Americas, lies an area known as the Mid-Atlantic Ridge. On the world map, the center of this ridge is indicated by what appears to be a stitched seam. In reality, this seam is a rift—a long crack in the earth's crust. In the South Atlantic, the rift is the boundary between the African Plate and the South American Plate. Many such rifts are found in the floors of the world's oceans.

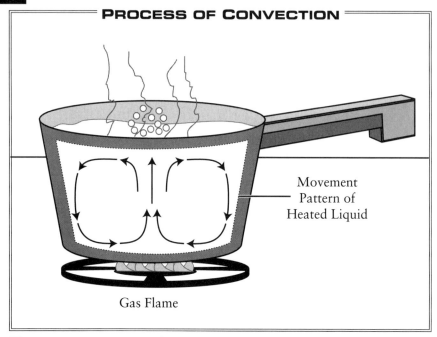

The convection process is demonstrated in a speeded-up and miniature form when a liquid such as soup is heated in a glass saucepan. The soup over the hottest part of the heating element rises quickly. When this very hot liquid reaches the top, it moves away from the continually rising column and begins to cool. When it has cooled further, it descends to the bottom of the pan. The process then begins again.

Liquid magma from the asthenosphere rises through a series of plumes toward the rift at the center the Mid-Atlantic Ridge. Most of the magma is trapped under the ocean crust, but some seeps up through the central crack in the crust and onto the floor of the ocean. Meanwhile, the magma trapped beneath the crust divides into two streams and flows away from the crack in opposite directions. The streams of magma are moved away from the rift by convection—the natural process by which very hot liquids move to a cooler location.

The streams of magma—flowing just beneath the ocean crust—act as liquid conveyor belts. The crust on either side of the rift is dragged away from the central crack. The small portion of magma that has not been trapped under the crust flows up

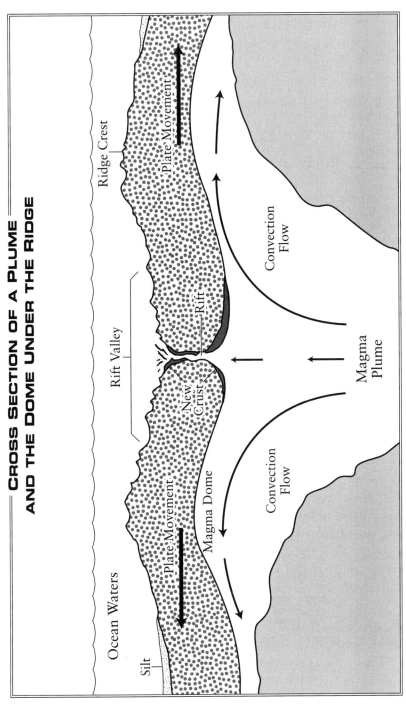

Convection currents in the magma help move the crustal plates. Molten rock fills the gap between the plates as they separate.

through the central crack and onto the ocean floor. It replaces the old crust as it is carried away by the conveyor-belt movement of the trapped magma.

The continuous process of convection that takes place under the Mid-Atlantic Ridge is duplicated at other ridge structures throughout the world. The amount of energy generated worldwide by this process is equal to the power of thousands of volcanoes, all erupting at the same time. The force is sufficient to move—at about an inch (a couple of centimeters) a year—the earth's crustal plates. This is a truly tremendous task because each of these plates is thousands of square miles in size. During the long periods of time of the earth's history, the force of the molten magma has slowly changed the location and appearance of every continent and ocean.

Although major changes in the earth's crust result from the process of convection at ridge-rifts, other important forces are also at work. The heat from the depths of the asthenosphere is slightly greater at some sections of a rift than at others. This

Fault Structure: The Great Rift Valley of Africa. The sections to the left and right in the photograph are slowly separating and the sea will someday fill the valley. (Courtesy of the National Geographic Society)

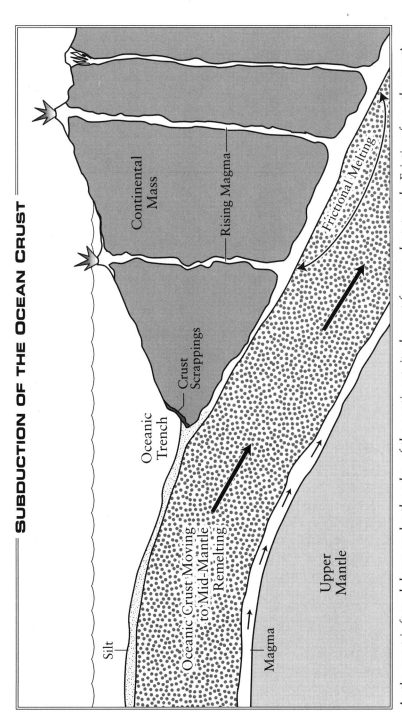

As the crust is forced down under the edge of the continent, its descent forms a deep trench. Friction from the crust's movement generates enough heat to melt rock. When this molten rock rises to the surface, it can cause volcanoes to form, such as Mt. St. Helens.

means that some segments of the crust will move slightly faster than the adjacent segments. The different speeds along the ridge are the cause of new cracks—called faults—which appear in the crust at right angles to the rift line. These faults extend on either side of the ridge line—sometimes for hundreds of miles.

Scientists now agree that new crust has been forming for hundreds of millions of years at the center of each ridge-rift. These rifts are known as "productive zones." The scientists also agree that the earth is not increasing in size. Therefore, something dramatic must be happening, somewhere, to old crustal material. Indeed, something is happening! Crust that is old and cold is reclaimed and recycled. This process occurs in deep sea trenches that are called "subductive zones." These zones, like ridge-rifts, sometimes can be found in the middle of an ocean. However, most of the trenches occur near a continental boundary.

Such a feature is seen at the boundary of the large South American Plate and the Nazca Plate—a much smaller oceanic plate. The South American Plate begins at the South Atlantic Ridge-Rift as ocean crust and extends thousands of miles to the west to include all of the continent of South America. Just offshore of the Pacific coastline of South America the westward-moving bulk of the South American Plate meets with the eastward-moving Nazca Plate. The Nazca Plate is forced downward by the larger plate. At this juncture, a deep ocean trench is formed—the most evident geological feature of a subductive zone. Such a trench can be as much as 5 miles (8 km) deep.

As the leading edge of the smaller Nazca Plate descends under the South American Plate, the overlying plate scrapes sedimentary rock from the surface of the smaller plate. The pressure from their collision, the scrapings of sedimentary rock, and the accumulation of other oceanic crustal material are changing the contour of the Andes Mountains. These mountains, that range near the west coast of South America, are still growing.

While some of the sedimentary rock that covers the basaltic rock of the Nazca Plate can be scraped off to change the shape of western South America, the basaltic rock itself is more dramatically

This volcanic eruption in 1952 occurred near the deep trench where the Pacific Plate, moving westward, dives under the islands of Japan.
(Courtesy of the U.S. Geological Survey)

recycled. The descending Nazca Plate follows the lower contour of the continental crust and eventually sinks through the mantle to the asthenosphere. At a depth of 50 to 100 miles (80 to 160 km), the basaltic crust begins to remelt. Thus the old crust is being recycled in much the same way as used cans and bottles are. This general pattern of events takes place in all the deep oceanic trenches in other parts of the world.

More Structure

Most of the action described so far takes place underwater. What is happening to the areas of dry land? It is helpful to recall that there are two kinds of crust. The continents are composed mainly of lighter, granite rock. Scientist believe that the granite solidified early in the evolution of the earth. Indeed, the continental crust is known to be many times older than the oldest oceanic crust. The crust under the oceans is made mainly of hardened magma called basalt, a relatively heavy type of rock.

Because of their different compositions, it is possible to think of the lighter continents as riders on the oceanic crustal plates. Some geologists believe that the heavy oceanic crust actually extends under the continents. Others believe that all of each continent is granitic rock. Most agree that although the continents are made of lighter rock than is the oceanic crust, their much greater thickness increases their weight and causes them to bulge down into the mantle.

The Present Geological Cycle

The Present location of the continents is temporary. Using plate tectonics theory as a guide, one can attempt to reconstruct the past. About 200 million years ago, tectonic forces appear to have caused all the continents to join together to create a supercontinent. The impact of this meeting probably formed the Appalachian mountain chain on the eastern side of North America and other major geological features.

Consequently, the crust at the collision zones was mangled and fractured and made weaker than the rest of the huge continental mass. Beneath the supercontinent, the intense heat of the core caused plumes of magma to drive upward through the mantle. When weak crust lay above the plume, the upward movement of the rising magma caused a dome to form. Eventually, the dome popped open, and magma flowed out. The result was an enormous volcano.

Probably, several domes formed at about the same time. Fractures could have appeared in any weakened crustal areas

A PASSIVE MARGIN

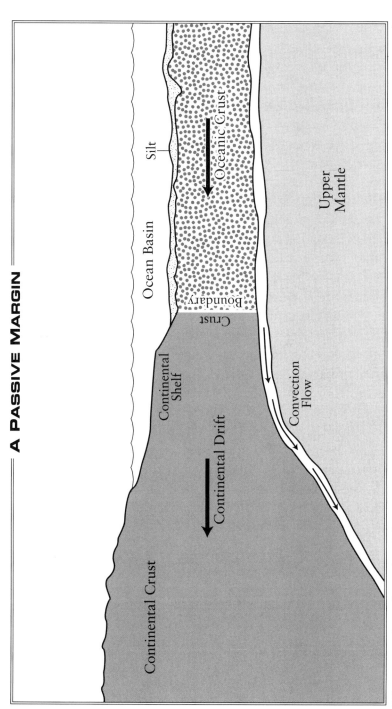

The eastern boundary of North America is a passive margin. It is not certain whether the oceanic crust continues under the continental crust or just abuts it. Either way, the whole mass of the ocean crust and the continental crust moves slowly northwestward.

between the domes. Scientists now think that the boundaries between Europe and Africa to the east, and North and South America to the west were formed by a series of faults—weak spots in the crust—that occurred between the curving series of domes. Therefore, the new plate boundaries were made by a sort of connect-the-dots process.

The magma that oozed out at the new domes and along the new fault lines was heavier than the older crust. Where the new crust piled up at the fault lines, the additional weight slowly caused the old crust to sink below sea level. Eventually, the sunken areas along the fault lines filled with water. Long, narrow oceanic inlets formed and then joined together. These inlets became the infant North and South Atlantic oceans.

Millions of years later, magma continues to stream out of the central rift in the mid-Atlantic, where it replaces the crust that has moved from the center. The new crust has gradually added to the distance between the coasts of Africa and the Americas. The result is known as seafloor spreading. By now, the east coasts of North and South America are a long distance from the active zone where the new crust is forming. These coasts are called passive margins. Earthquakes are rare, and there are no active volcanoes.

Around the Pacific rim are some interesting locations where three plates come together. Each plate is moving in a somewhat different direction. These so-called triple junctions are easily identified because they are the sites of many earthquakes and active volcanoes. Indonesia, Japan, and the south-facing coast of Alaska are representative sites of triple junctions. They are also relatively dangerous places to live because of the geological hazards.

Greenland, in the far North Atlantic, is another example of an unusual tectonic process. It appears that millions of years ago, the island occupied its own separate crustal plate. Although Greenland is now a part of the North American Plate, a rift on its western flank had once forced its separation from the mainland. Now, that rift is no longer active, and the island moves westward like the rest of North America.

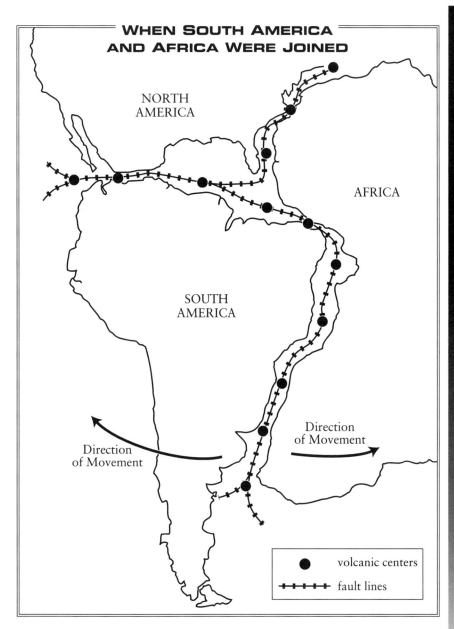

If South America and Africa were once joined, it is likely that they separated along a line of massive volcanoes. The faults in the crust that linked the volcanoes formed the rift that now occupies the middle of the Atlantic Ocean.

Iceland, just east of Greenland, is another particularly interesting piece of geography. It is one of the few places on earth where a ridge-rift seam shows itself above the surface of the ocean. Iceland's active volcanoes are a visible reminder of what is happening along the central seam of the Mid-Atlantic Ridge. South of Iceland, this volcanism, or volcanic activity, continues underwater for thousands of miles until the Mid-Atlantic Ridge joins the ridge system of the Antarctic Plate.

Research Methods

To gain information about the earth, geologists use research methods and information gathering techniques from all the sciences. A good geologist must be a capable physicist, chemist, and biologist. In recent years, many geologists have also become adept with mathematics—particularly high-level geometry.

The typical geologist not only needs to be knowledgeable about a range of scientific topics, but also must shift among three work settings—the office, the laboratory, and the field.

In recent years, an office has become a very important workplace for many scientists because of the increased use of personal computers. In addition to providing rapid mathematical calculations, the computer provides a way to gain knowledge on almost any scientific topic. It can be used to tie facts together into a reasonable and understandable pattern. In addition, computers are used to create communication networks among scientists.

The popular image of scientific research is centered on the laboratory. Moviemakers seem fond of scenes with electric spark generators and complicated glassware in which colored liquids bubble and fume. No doubt such images are good drama, but, at best, they are out of date.

In a modern laboratory, the scientist must exert complete control over each experiment in progress. The laboratory scientist also is expected to follow a tightly specified plan. Even when the work is guided only by curiosity, adhering to the plan is paramount. This care is reflected by the detailed records of each experiment. These records—kept in a diary or logbook—allow the researcher to study failures as well as successful experiments. Both kinds of outcome can be the source of new ideas.

Fieldwork has a much different atmosphere. The field researcher is an explorer who seeks new information in many different ways. Sometimes, the explorer travels to areas where no one has ever gone. At other times, the explorer examines some object that no one has ever studied. Also, the explorer might use new instruments to make revealing observations of previously studied features.

While working as an explorer, the field researcher must be adept at tolerating adverse conditions such as unpredictable weather, difficult terrain, and hostile people. Even political and social factors can interfere with the field-worker's plans.

Collecting the Facts

Geology is divided into several branches. The branches of geology that made large contributions to the study of plate tectonics were geophysics and oceanography. Sampling the ocean floor, a procedure used by scientists in both special fields, is a basic research technique. Scientists secure samples—called plugs or cores—of sand, rock, and other materials that make up the layers of the sea bottom. Plugs are obtained by driving a long pipe or sleeve into the bottom of the ocean. Such cores show the layers of material that were laid down over long periods of time. Many of the layers contain small shells that are the remains of animals that lived centuries before. Biological science is used to identify and date these marine animals. By examining each layer, scientists can speculate intelligently about the ancient oceans.

Echo sounding is another basic technique used by geophysicists. In this technique, sound waves—reflected by the layers of rock—are carefully analyzed using complicated mathematical

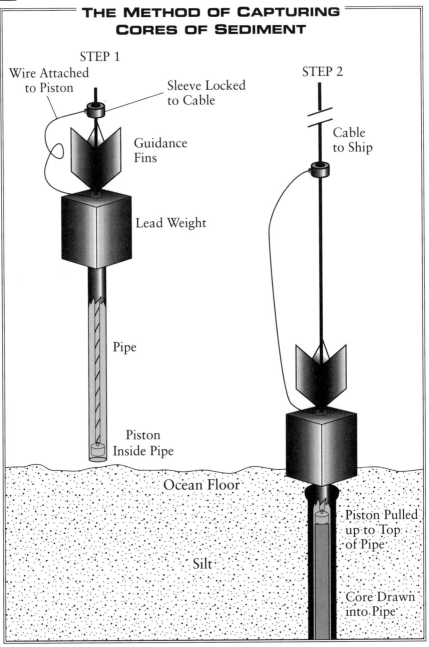

The hollow pipe is weighted with lead so that it will drop into the sediment with enough force to push the sediment into the pipe. A piston inside the pipe is withdrawn up the pipe to provide suction.

procedures. Useful sound waves come from natural sources such as the small earthquakes that occur daily all over the world. Other sound waves come from human sources such as the electronic pings of sonar or the boom of explosions. To aid their research, geologists purposely set off explosions on the surface of the earth or in underground holes. With the help of echo analysis, the reflected sounds can provide accurate information about the structure of the rock layers beyond the reach of the coring method.

In addition to describing the ocean floor, geophysicists describe the physical conditions of the waters of the earth's seas and oceans. Among their projects, they study the temperature and salt content of the water at various depths and locations. They also conduct surveys to determine if there are

Recovery of a Core: This photograph shows that the lead weight on the corer is about the size of a trash barrel. This particular core was taken in the Gulf of Alaska. (Courtesy of the U.S. Geological Survey)

SEEING WITH SOUND WAVES

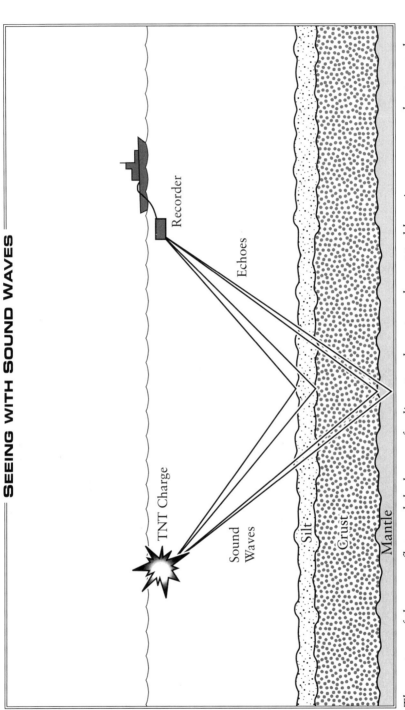

The contour of the ocean floor and the layers of sediment and crust can be mapped by using sonar sound waves and also by using the sound waves from exploding blocks of TNT.

changes in the strength of gravity from one part of an ocean to another.

Magnetism

The study of magnetism is yet another important part of geophysics. Anyone who has ever used a compass has seen the earth's magnetic field at work. Strange as it may seem, the earth's magnetic field is probably ten times weaker than that generated by a small battery and a few twists of wire around a nail. Given the weakness of the earth's magnetic field, it is remarkable that tiny amounts of this magnetism are found in ordinary rocks.

One usually thinks of magnetism as being a property of certain pure metals such as iron. However, many kinds of rock—such as hardened magma or lava—are magnetic because they contain small amounts of iron and other magnetic metals. As the once molten rock began to cool, it acquired traces of magnetism from the earth's magnetic field. As the rocks cool further, the magnetism is locked into them.

One of the great mysteries of earth science is the apparent wandering of the earth's two magnetic poles—the locations of maximum magnetism. The magnetic poles are the locations where the lines of magnetic force come together as they enter or leave the surface of the earth. Geophysicists can trace the apparent continuous path of the wandering poles by studying the magnetic properties of rocks that hardened at different times. Attempts to explain why and how the magnetic poles moved during geological time led some scientists to rethink the idea of continental drift. Such rethinking brought about the newer concept of seafloor spreading. This was a major step toward developing the theory of plate tectonics.

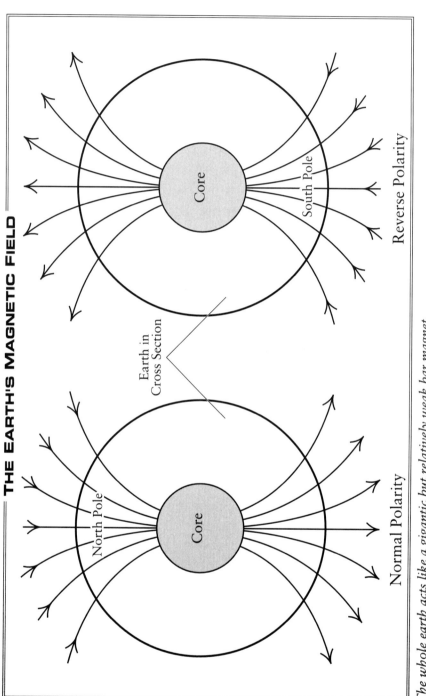

The whole earth acts like a gigantic but relatively weak bar magnet.

Measuring Age

Geologists can gauge the age of rocks in a variety of ways. For example, if the rock contains tiny fossils from creatures that lived at a particular time, the geologist knows that the rock was formed at approximately the same time. However, such methods are inexact. Both the date when the creatures lived and the time when the rock hardened are estimates. Scientists wanted very precise measurements and were not satisfied with a dating method that relied on fossils. In recent years, the study of nuclear science has led to the development of more accurate techniques for dating all materials.

Nuclear science was born in the 1930s, when scientists began to understand radioactivity. Radioactivity is the result of a breakdown in the composition of an atom. Particles that were part of the atom are released and flow outward at high speed. They found that some of the atoms in rocks changed over long periods of time. These changes occur in the internal composition of some atoms found in rock. These atoms give off radiation and become less massive. Using this information, scientists devised a mathematical formula to date rock samples. The complicated formula included computing the proportion of changed atoms to unchanged atoms in a rock sample.

The dating technique is demanding because it is difficult to separate the different types of atoms in the sample. Calculating the quantity of each type of atom is also an arduous task. Fortunately, in the 1950s, techniques to solve these problems were being developed in several geophysics laboratories. When rock dating was first used in conjunction with the study of rock magnetism, scientists realized that the particular magnetism of rocks depended on when the rock was formed. The researchers found that some young rocks had a magnetic orientation toward the present location of the earth's magnetic poles. Other rocks pointed to a location several hundred miles away from the present location of the earth's magnetic poles. They also found rocks whose magnetic fields pointed in exactly the opposite direction of the present locations. In other words, it appeared as if the magnetic field of the earth moved over time and sometimes completely reversed itself. It also looked as if the reversal had happened not just once but several times in the most recent 10-million-year period.

Scientists were dumbfounded by this new information. How could such an enormous event take place? What caused it? Will it happen again? If it did happen again, what would it mean to human life on the planet? These were some of the questions that had to be answered. The theory of plate tectonics can be regarded as a very important by-product of the search for these answers.

2

The Science Begins

There is no precise beginning date for the science of geology. For most of recorded history, philosophers—such as Aristotle in fourth century B.C. Greece—commented on unusual observations about the earth. Leonardo da Vinci—the great Italian genius of the late 1400s and early 1500s—left sketches and notes about marine fossil shells that he saw in mountain rock. However, none of the early thinkers attempted to weave together a comprehensive understanding of the history of the planet.

A Scots physician, James Hutton, was the first to organize his inquiries about the earth into a well-structured system. In 1785, he found the fossilized remains of sea creatures in the rock face of the Scottish mountains. Here, hundreds of feet above sea level and miles from the nearest sea were the shells of tiny marine creatures. How did they get there? Some said that these finds were proof of the biblical story of Noah and the worldwide flood. Skeptics, however, said that the biblical time frame could not fit the facts. According to the Bible, the flood lasted for a few years during Noah's lifetime. Hutton's observations revealed that the marine fossils were deposited in separated layers with the simplest marine creatures in the lowest layers. Therefore, he reasoned, the rock that made up the Scottish mountains had been under the sea for a very long period of time. It seemed clear to Hutton that it would have taken many thousands of years to lay down the layers of silt that contained the fossils. Many more centuries would have been needed to solidify the claylike silt into rock. Then, once the rock had formed, something had to push the rock layers high above the sea to build the mountains.

Hutton and others observed that the striations on the rock face—lines or bands that indicated each layer of silt—were often tilted at sharp angles. Such tilting suggested the work of titanic forces—such as a violent earthquake. No flood could cause such tilting and folding of layers of solid rock that are hundreds of feet thick.

Since observers had long noted that marine fossils were frequently found on dry land, it may seem strange that Hutton, in the 1700s, was the first to develop a grasp of this fact. At this time, it had become fashionable to be well versed in science. In order to maintain their positions as community leaders, some individuals wanted to demonstrate publicly their knowledge of

Scapegoat Mountain in Arizona: The layers of sedimentary rock that are exposed in this cliff face are similar to the layers of sediment that cover the ocean floor. (Courtesy of the U.S. Geological Survey)

scientific matters—so Hutton had a well-motivated audience. Also, Hutton was helped by an English writer named John Playfair. Playfair was able to make Hutton's ideas understandable and acceptable to the large numbers of people who had no formal scientific training.

Hutton's presentations were soon followed by observations from other interested Europeans. All the new information seemed to confirm a very basic fact. The surface of the earth could move up and down! When people thought about this, they were not too startled. They realized that it was not uncommon to observe fairly rapid subsidence, or sinking, of their own land. It seemed reasonable that if the earth could sink, it might also rise. However, even after these ideas were accepted, the cause of such processes remained a mystery.

Local change in the contour of the land was a minor issue compared to determining the age of the planet. While seeking the solution to this question, Hutton proposed that only observable processes such as erosion and volcanism would be employed in the calculations. This limitation prevented the introduction of religion and prophecy into scientific discourse. Using these guidelines, scholars were able to determine that the earth was much, much older than anyone had realized.

Hutton's theory of geology was based on the supposition that the forces that act on the earth were uniform over time. This theory generated confidence in the proposition that a particular kind of geological event might reoccur at several different places. For example, if diamonds were found in an unusual kind of bluish clay in one area, diamonds might be found in the same kind of clay elsewhere.

Such ideas were promoted effectively by another Scotsman, Charles Lyell. Lyell was the first geologist to compile a standard geology textbook. His book, published in 1830, was revised many times before being replaced by a newer text in 1876.

During Hutton's and Lyell's time, the earth was believed to be about 8,000 years old. The estimate was based on biblical information about the number of generations between the birth of Adam and Eve and the birth of Jesus. For most people, 8,000 years seemed long enough for everything—both prehistoric and historic—to have occurred.

Hutton's principles and Lyell's studies led investigators to consider other, non-biblical sources of information about the age of the earth. New estimates were based on the average temperature of the earth's surface. If the planet had originally been all molten rock, one could determine its age by calculating the time needed to cool and solidify its surface. Such computations led the English scientist Lord Kelvin to propose that the earth was more than 100 million years old. Scientists now know that this figure was still far too small. However, the date was not corrected until about 100 years later when scientists learned that radioactivity produces heat. When this information was added to the calculations, the earth was found to be much, much older.

Most modern methods of calculation suggest that Earth is about 4.5 billion years old. Fossil evidence suggests that life has existed on Earth for more than half that time. Recorded human history—about 7,000 years—is equal to one small tick of the celestial clock.

The Origins of Mountains

After Hutton and Lyell established the fundamental concepts of systematic geology, both the theory and practical application of the science developed rapidly. Scientists sought a theory to explain how the general profile of the land was formed and, in particular, how mountains were built.

A few brave souls came forward with the notion that the earth had expanded and that the continents had been pulled apart. The analogy of a balloon being blown up was used to make this idea more understandable to general audiences. It was argued that the expansion process provided the force that built the mountains from a flat surface. However, few people accepted this idea.

A slightly more plausible notion was soon advanced. Some said that the earth had shrunk in size over geologic time. Since many believed that the earth had begun as a hot, molten ball, it seemed possible that it had become smaller as it cooled. People knew that ordinary metal objects do shrink when they cool. For example, cooks had observed that the metal lids used when canning food become slightly smaller—and therefore tighter—as they cool.

Another analogy was sometimes used. The earth's shrinking from whatever cause can be compared to an apple drying. As the apple dries, its skin begins to wrinkle. This process provided a graphic but incorrect idea to explain the rise of mountain chains and the subsidence of soft spots to form the ocean basins. However, a nice image is not a scientific explanation.

By the end of the 1800s, several interesting notions about the process of mountain building had been advanced. However, the process was not explained satisfactorily until the scientific revolution of the 1960s.

In the early 1900s, volcanism and earthquakes were used to explain some of the observable conditions of mountain building. For example, the earthquakes that took place at the San Andreas Fault in California were thought to provide the forces that generated the sidewise movements of the earth's crust. It is now known that this is incorrect. These lateral movements—deep underground—generate the pressures that are *relieved* by earthquakes.

Early in this century, the crust of the earth was considered a flexible skin. Therefore, if the skin of a particular region was forced downward by the weight of a glacier or layers of silt and seashells, the surrounding area would be forced to rise. Likewise, if the layer of soil became deeper because of silt deposits from erosion, the thick, heavy surface sank a little. Even with such sinking, however, the area of added thickness usually had a higher profile than the surrounding, thinner crust. Skeptics had problems with such explanations of mountain building, and theoreticians generally sidestepped this basic question.

Because of the pioneering work of many geologists, it was clear that correlations existed between the presence of certain kinds of geological formations and the presence of valuable minerals, gems, and petroleum. Prospecting and mining companies soon became more scientific about their explorations. In the 20th century, geological knowledge was vital to the development of modern oil fields and mining sites. Geologist's knowledge of the composition and structures of rock was soon complemented by new methods of obtaining information. One such useful method was based on sending strong, sharp sound waves down into the earth and recording their echoes. The use of this technique became crucial in the further study of the formation of the earth.

3

The Visionary and the Skeptics

The founders of geology and the related earth sciences gave little thought to such seemingly irrelevant matters as the apparent fit between the contours of the west coast of Africa and the east coast of South America. After more accurate maps began to appear in the mid-1500s, some geographers noted the parallel outlines of the two continents. Around 1620, Francis Bacon, an English lawyer and philosopher, wanted to encourage practical scientific research. Bacon wrote briefly about the peculiar match of the coastlines. However, neither Bacon nor those who followed could bring themselves to believe—or even conceive—that the two continents had ever been joined along the length of their respective coastlines. Many years later, people observed that the Atlantic coastlines of Africa and South America fit together like pieces of a jigsaw puzzle. Speculation based on the apparent fit was given wider publicity by cranks and attention-seekers. However, few professional geographers or geologists paid serious attention to such a farfetched idea.

Solid evidence that the continents might have been connected in the distant past began to surface in the mid-1800s. Geological structures—such as long veins of a particular type of rock—were observed to end at the western shoreline of Africa, only to reappear across the South Atlantic Ocean on the eastern shore of South America. Likewise, certain fossils found near the African coast could also be found in South America. Some reasonable people began to believe that these discoveries were more than just coincidence.

For years, however, the most popular explanation was that an extensive land bridge had connected Africa and South America. Some people believed that this land bridge was ripped from the earth by a passing comet. They suggested that the rocks ripped out by the comet eventually formed the moon. If that amount of rock had been sent into space, a deep trench would have been created in the primitive Atlantic Ocean, and waters from other primitive oceans would have rushed into the void. Shallow seas would have been drained of their water, and low-lying areas would have been left dry. This was a story fit for telling around the fireplace on a cold winter night, but not a sound scientific theory.

A Turning Point

German scientist Alfred Wegener was the first person with impressive scientific credentials to consider the present location of the continents as a major scientific problem. In 1911, he theorized that the continents and all other areas of dry land had once been joined together as a single gigantic continent. Wegener thought that seams, like the cracks in a large sheet of ice, had appeared in the natural rock surface. Such seams defined the boundaries of the new, smaller continents. The pieces of the once unified land mass began to drift apart. Over a long period of time, they came to their present locations. Indeed, Wegener theorized that they were still moving!

YOUNG ALFRED

Alfred Wegener was born in Berlin, Germany, in 1880. His father was a scholarly Protestant minister, and his grandfather, great-grandfather, and great-great grandfather had all served the church as clergymen. Such service was a family tradition that went back for more than 100 years. Alfred and his younger brother, Kurt, broke the pattern. They, too, were scholars, but were not called to a religious life. Instead, they were bound for a life of adventure.

When Alfred was taking his advanced university degree in astronomy and physics, he and Kurt became interested in free ballooning. The Wegener brothers soon set an endurance record of 52 hours aloft. In 1906, they entered an important international contest for balloonists that drew 17 contestants and much publicity. They did well in such competitions and often persuaded their young collegiate friends, both women and men, to join them for a ride in the balloon's gondola.

THE RESEARCH CAREER BEGINS

Alfred Wegener received his doctorate in the spring of 1906 and began his scientific career as a meteorologist. Given his hobby of going aloft in balloons, it was fitting that he was employed by the Royal Prussian Aeronautical Observatory near Berlin to study conditions in the upper atmosphere. However, the job did not last very long. After only a few months, Wegener was invited to join a scientific expedition to Greenland. Perhaps his invitation stemmed from his fame as an adventurous balloonist rather than his good research credentials.

The research team of 28 scientists was led by the Danish scientist Ludvig Mylius-Erichsen. By the spring of 1907, the team had established a base camp on the northeast coast of the huge island. In the early summer, after the supplies were organized, Mylius-Erichsen and two colleagues set out on a modest scouting expedition. When the small group was some miles from the base camp, they were caught in a bad storm. They could not move for days and ran out of food. All three members of the scouting party died.

The whole expedition was called off, and all the surviving scientists returned to Europe. After arriving home, Wegener accepted a teaching position at the University of Marburg. While there, he began to give serious attention to his loosely formed ideas about continental drift. These ideas had grown out of his basic interest in geography. He wrote two brief papers on his speculations, which were published in a German magazine for professional geographers.

In spite of this scholarly success, Wegener was not content to pursue such theoretical ideas to the exclusion of other scientific projects. He was eager to return to Greenland and continue the study of its climate. The tragic events of his first journey of scientific exploration had not dimmed these ambitions. In 1912, Wegener and Johann Peter Koch, a colleague he had met on his first trip, planned a new expedition. Koch and Wegener were to be the only scientific staff members for this project. They expected to hire two local residents to serve as both guides and porters when they arrived in Greenland again.

They scheduled the trip in two stages. The first stage consisted of traveling to Iceland and stopping over for 20 days. Koch wanted to use Icelandic ponies as pack animals for the main expedition. He had wisely decided to test his idea in the relatively mild weather in Iceland before exposing the animals to the really harsh conditions on Greenland. Both men were pleased to see that the ponies were able to do the work.

Koch and Wegener had made preliminary arrangements to land at an Inuit village about halfway up the east coast of Greenland. They arrived with the ponies and five large sleds to haul their supplies. Two of the villagers were waiting to guide them and help handle the equipment.

The second stage of the plan was a trip across the heart of Greenland. First, they established a base camp on the crest of ice above the Inuit village where they had landed. After spending the winter on the crest, they planned to travel across the island on a diagonal southwesterly course of about 700 miles (1,120 km). This portion of the second stage would take the explorers over the central ice dome to another Inuit village on the southwest coast.

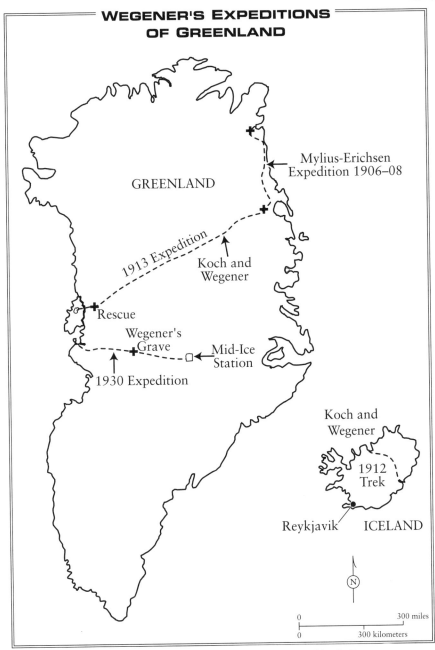

Wegener's Expeditions on the Island of Greenland: Alfred Wegener was fascinated by Greenland and participated in three expeditions to explore its secrets.

It was already winter when they pulled their gear up the steep slope of the ice dome and made their base camp on the top. They were the first explorers to establish a base camp on the Greenland icecap and live there for the entire winter. When the terrible winter storms eased in the spring of 1913, the exploration team began the trek southwestward across the island.

Except for some narrow areas along the coasts, Greenland has a nearly complete cover of glacial ice. A major scientific goal of the expedition was to determine the thickness of the ice shield near the center of the island. Using echo sounding, they found that this ice shield was almost 2 miles (3.2 km) thick—it penetrated downward below the level of the sea.

The two scientists and their two guides had covered about 670 miles (1,072 km) of their trek when the last of their ponies died of exhaustion. Fortunately, Inuits found the explorers when they were still some distance from their final destination. The four men were down to their last few pounds of food when they were rescued. They had taken great risks to make their scientific observations and were lucky to be alive.

Life at Home

As soon as Wegener returned to Germany in the fall of 1913, he married Else Koppen, a young woman he had first met in 1908. They had a short honeymoon and by the beginning of the school term, Wegener resumed teaching at the University of Marburg, the position he had accepted in 1907. However, his quiet life was soon interrupted by the outbreak of World War I in the late summer of 1914. Wegener was a reserve officer in the German Army and was immediately called to serve. When the German Army invaded Belgium, Wegener was wounded in the arm. That wound soon healed, and he quickly returned to the front. His next mishap was a serious neck injury that required a long convalescence. This wound ended his military career. He now had time to renew his thoughts about the theory of continental drift.

The papers he had written before the war did not include all the points he wanted to make. He decided to focus on the task of putting all his ideas into a well-organized presentation. The short book that resulted from his efforts was called *The Origins of the Continents and Oceans*. It was first published in 1915 while the war was still in progress. Unfortunately, the disruption of communications caused by the war did not allow his book to be widely circulated to the international scientific community.

A New Position

Wegener's wife, Else, was the daughter of Vladimir Koppen, a meteorologist of considerable prominence. He was the chief weather analyst at the German Marine Observatory in Hamburg. Koppen was 72 years old in 1919 and ready to retire. He arranged for his son-in-law to become his replacement.

The older man retired from his job, but did not stop thinking and working on his scholarly pursuits. Koppen and Wegener entered into a highly productive collaboration on a wide range of scientific topics. However, Wegener still reserved part of his time to pursue the theory of continental drift. He had already revised his book in 1920, and the third version, published in 1922, was translated into English.

In 1924, the whole family—Wegener, his wife, both her elderly parents, and the Wegener's three children—moved to the town of Graz in Austria, where Wegener had been appointed to a professorship at the university. His collaboration with his father-in-law had flourished, and one result was a book called *The Climates of the Geological Past*. This book included a discussion of the apparent movement of the magnetic North Pole and the discovery of tropical fossils in arctic locations—important topics in the theory of continental drift.

THE RISING DISPUTE

Since the 1922 version of Wegener's book on continental drift had been translated into English, it became much more accessible to professional geologists around the world. The ideas were definitely controversial. Many geologists protested that Wegener's notions were too outlandish to be given serious consideration. However, it was clear that Wegener had put his finger on some inescapable and unexplainable facts.

The presence of glacial remains in the rocks of South Africa was one such fact. The existing theories about the origins of the continents were clearly inadequate, and some revisions were required. Were Wegener's ideas about moving continents an acceptable alternative? This question had to be addressed.

In 1926, as part of the annual meeting of the American Association of Petroleum Geologists, a special symposium was held to discuss Wegener's ideas. Aside from the person who chaired the special meeting—the Dutch vice-president of an American oil company—most of those in attendance were strongly opposed to the idea of continental drift. The chair argued for open-mindedness and fairness. However, most of the more senior geologists, such as Professor Rollin Chamberlin from the University of Chicago, asserted that the theory was totally unscientific. Other major figures from the United States and the United Kingdom were even more harsh and personal in their attacks on Wegener's ideas.

Indeed, the theory of continental drift was weak on two major points. First, Wegener saw the continents as virtually independent bodies. This meant that the continents, in order to move, would have to plow through the crust at the bottom of the oceans. Any such motion, no matter how gradual, would leave enormous scars behind. The movement of each continent would have left a trail in the form of broken and displaced crust. No physical evidence of the required sort could be found.

The second weakness was the lack of a believable energy source. All geologists and almost everyone else recognized that the inner earth was hot because hot lava flows from volcanoes.

However, no one could see how this source of energy might work to move the continents.

Wegener proposed that slight changes in the tilt of the earth's axis of spin would tend to propel the continents. Again, any such wobble would have remarkable side effects over relatively short time periods, and none were detectable.

Scientists who were sympathetic to Wegener and his ideas suggested that a theory could be useful to science even if it contained some inconsistencies. The thrust of the argument in Wegener's favor was that his ideas should be the basis for further research. Some hoped that the ideas would energize both those who agreed and those who disagreed with the concept of moving continents. Both groups could contribute to scientific advancement by their search for new evidence. However, the skeptics remained so negative that they would not discuss the idea further. In fact, some tried to prevent their students from even hearing about the theory.

Wegener did not help his cause when he made one terrible measurement error. He claimed that he had evidence that Greenland had moved westward at the rate of almost 120 feet (36 m) in a single year. This figure was incredible to most scientists and, in fact, was wrong. His Danish friends had provided geographic measurements that were based on the standard navigational techniques of 1907—the year of Mylius-Erichsen's ill-fated expedition. The measures were full of errors, but because they supported his ideas, Wegener accepted them. This led to ridicule by senior geologists around the world. Unfortunately, this ludicrous claim killed any interest by serious scientists in his theory. Some argued later that because Wegener was German rather than English or American and a physicist and meteorologist rather than a geologist, his cause would have been lost even with a much better base of facts.

THE FINAL EXPEDITION

After Wegener lost the debate about his theory, his standing as a geologist was in question. However, he still had his position at a fine university, a loving family, and loyal, professional friends back home. One of these friends, Johannes Georgi, was a former student. By using Wegener's techniques for meteorological observations—such as tethered balloons—Georgi discovered what is now called the jet stream. The jet stream is a high-altitude river of air flowing constantly from west to east in the northern hemisphere.

Georgi's initial observations had been made from a weather station on the northwest part of Iceland. Georgi deduced that this stream of air might influence weather in Europe. He was confident that observations further to the west, on Greenland for example, might be very valuable for making long-range weather predictions. His ideas were logical and later proved to be correct. Consequently, Georgi was able to convince Wegener and others that an international effort should be made to establish a weather station on the Greenland icecap.

After months of planning, an expedition was launched under Wegener's leadership. In the summer of 1930, a temporary weather station was set up in the middle of Greenland. Leaving Georgi to operate the station with one other man, Wegener and other members of the party returned to the base camp on the west coast of Greenland. Their immediate mission was to bring in equipment and supplies for the two station operators. Substantial resources would be needed if these station operators were to survive through the winter.

Bad luck plagued the supply effort. A party of 15 men was formed to transport the materials needed at the station. This party dwindled to three after trekking through outlandish cold for 100 miles (160 km). Twelve of the guides and porters abandoned the effort and returned to the base camp on the west coast. Wegener felt duty-bound to go on.

The reduced group of three, two scientists and one Inuit guide, finally arrived at the station. However, they had been forced to abandon much of the material with which they had started. Fritz Loewe, the other scientist in Wegener's party, had such badly frostbitten feet that he could go no farther. He decided to spend the winter months with Georgi and the other station worker at the crude facility. Even with the supplies that Wegener and Loewe had brought, there was not enough to support all five men during the long, cold winter. The next day, Wegener and the guide headed back for the coast with their worn-out dogs pulling a sled. They did not make it. When summer finally came to Greenland, Wegener's body was discovered on the trail. He

Alfred Wegener was one of the group of explorer/scientists that includes such figures as Marco Polo, Charles Darwin, and Admiral Byrd. (Courtesy of Sigrid Berge)

had died of a heart attack at 50 years of age. The guide's body was never recovered.

It would be another 36 years before Wegener's theories were partially vindicated. He was a visionary who saw a great truth, but had insufficient evidence to prove it. Others would step into the breach and collect the evidence that Wegener lacked.

4

The New Earth Sciences

*I*n the late 1920s and early 1930s, the study of geology looked as if it were running out of steam. Geology was still important to prospectors in the oil business and builders seeking firm foundations for tall buildings. However, the field of geological research was becoming rather boring to many of the specialists. In the United States, lengthy surveys of the earth's surface continued to be sponsored by individual states. Sadly, once a survey was completed, there was little more of interest or significance for a geologist.

During this period, most geologists gave Wegener's ideas little further thought. If they wrote about continental drift, it was to scoff and make fun of the idea. However, Wegener's ideas never fell into total obscurity, partially because they were easily made available. The ideas seemed to intrigue essayists and readers of popular magazines. During the 1930s, one could expect an article or two each year in such sources as *American Mercury*, *Literary Digest*, and *Scientific Monthly*. In 1941, there was even a piece in *Time* magazine. Not all these articles were supportive but, year after year, people were aware that the idea of continental drift was still alive.

Oceanic Geology

Many traditional geologists—whether or not they agreed with Wegener—were determined to revitalize their science. These scientists all believed that research on standard geology should go forward. However, some believed that this research, although very important, would not solve their basic theoretical questions in a timely manner. This small group of people thought that the most important answers were to be found underwater, particularly on or under the deep ocean floor. In the 1930s, Professor Richard M. Field of Princeton University was a strong believer in this idea.

Many geologists disagreed with Field's conviction that the sea would become the new frontier of scientific research. Their response was "easy to say but hard to do." Although some others agreed with Professor Field's beliefs, they saw great difficulty in actually doing the work and asked many practical and probing questions. How can one make accurate observations in the dark, cold, ocean depths, where water pressure can crush a submarine like an eggshell? Who pays for the expensive, special equipment and the ships to carry the scientists around the broad oceans?

To eliminate such problems, Field proposed to start his research program at the water's edge. He planned to expand the studies into deeper waters as the necessary technology improved. He hoped that the results of his shallow-water studies would motivate fellow geologists to begin additional—and costlier—research. In the mid-1930s, Field began his own preliminary studies on the shallow slopes off the Virgin Islands.

His research mission was soon halted by the outbreak of World War II in 1939. However, before that turning point, he had convinced two young scientists, Maurice Ewing and Harry Hess, that the ocean depths were the place to find crucial evidence about the nature of the earth. These two men became leaders in oceanographic studies during the postwar period. In different ways, each was strongly influential in the wide acceptance of continental drift theory.

Ewing began his research by doing echo soundings of the continental shelf, the area of shallow ocean just beyond the edges of the continents. His work was, at first, conducted off the beaches of New Jersey. Later, with help from Professor Field, he was invited to conduct his studies aboard one of the research ships belonging to the Woods Hole Oceanographic Research Center on Cape Cod, Massachusetts. When the ship was temporarily anchored a few miles from shore, Ewing began experimenting with echo sounding. Although he collected only a small amount of useful information, Ewing became familiar with using scientific instruments in the open sea. Because of his experience, the U.S. Navy began to support Ewing's work after the United States entered the war in late 1941.

Professor Field's other protégé, Harry Hess, was a postgraduate student in the geology department at Princeton. When he finished work on his doctoral degree, he left Princeton to activate his commission in the U.S. Navy. After the United States was drawn into World War II in December 1941, Hess was given command of an attack transport in the Pacific. He was engaged in U.S. Marine Corps assaults on four islands and ended the war with the rank of commander.

During the war, Hess did not give up the interests that Field had aroused. Whenever possible, he did bottom soundings with the ship's sonar equipment and located important features such as seamounts—mountains rising up from the bottom of the sea. Many of these mountains seemed odd because they had distinctively flat tops rather than the sharp peaks of continental mountains. During his limited free time, Hess tried to find an explanation for their peculiar shape.

During the war, Ewing and Hess were not the only ones interested in research that supported the national defense. Many of these scientists, like Field's protégés, were engaged in work that involved an investigation of the world's oceans. They were encouraged by military commanders who needed answers to fundamental questions related to submarine warfare. How could the British and Americans contain and defeat the threat of German submarines in the Atlantic? How could they mount effective submarine attacks on the Japanese Navy in the Pacific?

To solve these problems, scientists had to know more about ocean currents and the temperature and salt content of the layers of ocean water. It was necessary to improve their echo-location devices. These sonar systems sent out sound waves that had to penetrate through layers of ocean water before contacting and bouncing off enemy vessels.

The scientists also had to develop ways to give the submarine captain a clear picture of his ship's location in the ocean and the best course to navigate to and from his home base. In addition, new instruments were needed to apprise the captain of nearby hills and valleys on the ocean floor so that the ship could be hidden in case of an attack.

The need to solve these practical problems pushed the military leaders into supporting oceanographic research and submarine technology. The momentum that built up during the war continued afterward. The military leaders saw that the wartime research had improved combat effectiveness. The scientists realized that the effort directed at the solution of military problems had also yielded useful information about basic theoretical questions. Therefore, the improved techniques for making scientific observations served the needs of both the military and the scientific communities.

One outcome of this realization was the establishment of the U.S. Office of Naval Research. Top government officials believed that such organizations could help prevent the postwar breakup of productive research teams. To achieve this goal, scientists involved in oceanographic research were provided with money, equipment, and ships for their expeditions.

Some of the scientists involved in wartime oceanographic research remained as government employees. They were hired as civilians by government agencies such as the U.S. Coast and Geodetic Survey. Others resumed their academic careers in colleges, universities, or university-sponsored research centers. People who had shown outstanding abilities during the war were often given positions of administrative leadership in these various organizations. The scientists went on to define the research strategies that evolved during the 1950s and 1960s. During this period of revolutionary change, a focus of geology shifted from studying restricted areas on the continents to attempting to comprehend the planet as a whole.

New People in New Places

By 1945, many scientific research facilities in continental Europe and in Japan had been abandoned or damaged during the war. Financial support to rebuild and restaff such centers was slow in coming. In some countries, all funds were needed for emergencies such as the outbreak of epidemics or the prevention of famine. In contrast, the institutional bases for scientific research in Britain and North America were still intact. Many faculty members returned to their former positions and restarted their research projects. Some, like the geologists, returned with many new research ideas gained from their wartime experiences. Fortunately, government officials in Britain and North America were willing and able to provide money to support new projects.

At the same time, students whose education had been interrupted by the war were returning to the classroom. Many war veterans who had not thought of seeking a college education changed their minds when they became eligible for financial support from the government. These veterans, both young and not so young, began to enter college. Young people who had just completed high school were also ready to enter institutions of higher learning. The resultant flood of students overwhelmed some established institutions. Classrooms and laboratories were expended as quickly as possible to accommodate these large numbers of students. In some cases, small research facilities—once considered minor divisions of traditional academic departments—were given increased independence.

Two such research centers were particularly important to the revolution in the earth sciences. One was the geophysics unit at Columbia University in New York City. It remained part of the university but moved up the Hudson River and became the Lamont Geological Observatory. The other was the Scripps Institution of Oceanography, which affiliated with the University of California at San Diego.

Traditional academic departments of major universities also were involved in the study of underwater geology. Princeton University in New Jersey, the University of California at Berkeley, the University of Toronto in Canada, and Cambridge University in England were major centers of activity. Independent research organizations, such as the Woods Hole Oceanographic Institution in Massachusetts, provided large quantities of oceanographic data. In addition, the U.S. Geological Survey and other government institutions participated in the revolution. From the mid-1940s to the late 1960s, hundreds of scientists from dozens of research organizations contributed to the building of the new theory. However, Lamont, Scripps, Cambridge University, and the University of California at Berkeley were the home institutions of the key participants.

5
A Tough Man for a Tough Job

*F*ield research can be dangerous. Although Alfred Wegener's death on the Greenland glacier was not directly related to his work on the theory of continental drift, it carried a message about field research to his successors. Conducting observations in earth science could, at the very least, require scientists to forfeit their physical comfort from time to time.

The risks were not all physical. Wegener's story reveals that one's reputation as a legitimate scientist might be put on the line. Wegener had the advantage of a secure job outside the profession of geology. Criticism of his ideas might have hurt his pride, but his reputation as a scientist rested solidly on his work in meteorology and climatology. Professionally, he readily withstood the negative reaction to his concepts about continental drift.

The people who eventually revitalized Wegener's ideas were much more closely identified with geology. From a professional standpoint, they were more vulnerable than Wegener and had to be more cautious. While being cautious, they had to be tough and retain their self-confidence in the face of criticism and frustration. Maurice Ewing was willing to accept both the physical and psychological risks that a pioneer must endure.

The Early Years

Ewing, whose career would focus on underwater studies, was born and educated in the dry plains area of north Texas. In high school, he was good in mathematics, and this talent helped him gain admission to Rice Institute in Houston, Texas. He majored in physics during his undergraduate and graduate training and received his doctorate in 1930.

Ewing began his teaching and research career at Lehigh University in Bethlehem, Pennsylvania. In 1934, he was visited by Professor Richard Field from Princeton University and William Bowie from the U.S. Coast and Geodetic Survey. Field had heard of Ewing's studies using echoes to identify subsurface rock formations. To conduct his research, Ewing set off dynamite charges, which he had placed in small excavations. He used specialized recording devices to measure the resultant sound waves as they bounced back from hidden layers of rock. Although this work was important, the older men convinced Ewing that more significant information could be gained by using underwater sound waves to conduct research on the continental shelf.

The continental shelf extends outward from the seashore for several dozen miles. Over that distance, the ocean is relatively shallow. At the end of the shelf, however, the bottom drops off sharply. The men agreed that it would be safer and less expensive to conduct the initial research near the coast of Virginia. Scientists believed it likely that the continental shelf was composed mainly of sediments washed into the ocean. However, no one knew how these sediments were arranged or what the hidden continental foundation looked like. By analyzing underwater sound waves, they hoped to understand both the composition and appearance of this marine landscape.

To conduct his experiments, Ewing generated sound waves by setting off charges of TNT on the bottom of the ocean. The sounds and their echoes were recorded at a series of listening stations that also rested on the bottom. Ewing's first oceanographic projects did not produce very clear results, but Ewing never turned back. He had been captivated by the idea of studying the floor of the sea.

In the late 1930s, Ewing—with the help of the influential Professor Field—received the support of the senior people at Woods Hole Oceanographic Research Center. They allowed him to use their research ship, *Atlantis*, for a two-week cruise. He was able to expand his research capabilities by discharging blocks of TNT farther out at sea. The patterns of these sound waves were far more understandable. For the first time, he was able to map a small section of the ocean bottom.

After several more years at Lehigh University, Ewing transferred to Woods Hole in 1941, the year that the United States entered World War II. He spent the duration of the war doing defense work for the U.S. Navy at the research center. In 1947, he was offered a position at Columbia University and moved to New York City.

In the setting provided by Columbia, Ewing created and nurtured a major research organization. He began as a faculty member of the geology department, but was soon on the move. In 1949, he was appointed director of the newly established Lamont Geological Observatory. He and his team were relocated from their cramped quarters on the campus in Manhattan to an abandoned estate north of the city on the Hudson River. The property had been deeded to the university by the widow of Thomas Lamont.

In order to make the observatory self-supporting, Ewing set out to obtain research contracts and grants from government sources. This money, and a small endowment from the widow, provided research opportunities for a steady flow of graduate students. Twenty years later in early 1969, the Henry L. and Grace Doherty Charitable Foundation gave $7 million to Columbia University as an additional endowment for the observatory. At that time the name was changed to the Lamont-Doherty Geological Observatory.

The Program

Ewing's general research strategy was to go to sea whenever possible and then measure and record everything available to him. By using a variety of instruments—many designed or redesigned by himself and his colleagues—he measured magnetic fields, gravity differences, and the direction and speed of ocean currents. The depth of the ocean was measured by echoes from sonar-type equipment. In addition to these measurements, Ewing used charges of TNT to aid in his study of the layers of rock and sediment that lie below the bottom of the sea.

Ewing also stressed the importance of taking large numbers of cores. Cores—or plugs—are samples of the materials that make up the ocean floor. The samples are secured by driving a long pipe into the bottom of the ocean and then bringing the sample to the surface. The plugs, which can be 40 feet (12 m) long, reveal the layers of silt and other material that have been laid down over geologic time spans. The layers are like those in sedimentary rock that are exposed by an excavation on land. Ewing and his people became so adept at taking cores that the Lamont observatory soon had the world's largest collection of seafloor samples.

An even higher priority was given to the study of underwater sound waves. In addition to recording the sonar pings used to determine surface features on the seafloor, Ewing studied both synthetic and natural sources of sound. Sonar pings are generated by electronic machines. The other synthetic sounds that he measured were generated by exploding TNT underwater. He also recorded natural sounds that were generated by earthquakes. Minor earthquakes occur in oceanic regions almost every day, but their sounds can be difficult to interpret. However, such sounds provided Ewing with a steady source of information.

In 1947, shortly after moving to Columbia University, Ewing made his first eye-opening discovery. While conducting research from a ship provided by Woods Hole, Ewing determined from his sound wave observations that the ocean crust was much thinner than most experts had believed.

More Findings

Ewing's first extensive expedition into the far Atlantic took place in 1947. Ewing and his crew sailed Lamont's research vessel,

Atlantis, beyond the continental shelf. They observed a large, flat area nearly 3 miles (4.8 km) below the surface of the water. It was almost perfectly level except for a few minor hills and seamounts (underwater mountains). Ewing noted this interesting feature as they progressed eastward across the Atlantic. The flat appearance of the ocean floor showed little variation until they began to approach the far reaches of the Mid-Atlantic Ridge. The ridge, a range of low mountains on the sea bottom, had been discovered during the laying of the first submarine telegraph and telephone cables. At that time, the feature was assumed to be of no great geological significance. Indeed, the appearance of the ridge was far less interesting to Ewing and his crew than the totally flat sea bottom that they had observed. The Mid-Atlantic Ridge, however, would prove to be a very important factor in the theory of seafloor spreading.

The "abyssal plain"—as the area of level sea bottom is called—is composed of an even deposit of sediments. Such sediments are formed of claylike material made of shells from tiny marine animals. For millions of years, these shells and other materials have rained down from the near surface waters of the oceans. Ewing took cores of the sediment and, curiously, layers of sand were found in the samples. The sand was mainly in the form of ground-up quartz, which results from the erosion of continental rocks. This type of sand is carried by the water that flows to the ocean's edge from the drainage areas of streams and rivers. Therefore, this sand should be found on ocean beaches but not hundreds of miles from the shore in the ocean's deeps!

In 1949, by a quirk of fate, Ewing was placed in charge of a Woods Hole ship that was sailing in the Gulf Stream off the mid-Atlantic coast. He was given the opportunity to trace—by echo sounding—the course of the Hudson Submarine Canyon. This canyon was known to begin at the mouth of the Hudson River, near New York City. However, no one had explored its length. In the short time available to him, Ewing traced the canyon's path as far out to sea as possible. He did not state what he hoped to find. It seems reasonable, however, that he suspected a possible relationship between the mysterious sand found in the deeps and the geological formation represented by the Hudson Canyon.

As the year progressed, he continued an intermittent survey of the canyon. He soon found that seaward from the continental shelf, the canyon led into an underwater valley that resembles a deep, wide riverbed. About 100 miles (160 km) farther out, this valley is 3 miles (4.8 km) wide, 900 feet (270 m) deep, and lies below about 2 miles (3.2 km) of ocean water. Near the shoreline, where the Hudson Canyon cuts into the continental shelf, it forms a colossal gorge. The survey, completed in 1953, also showed that the valley continued for more than 1,000 miles (1,600 km) across the Atlantic.

The valley of the Hudson Canyon seemed to be the result of a massive mud slide or avalanche. Perhaps a mass of material was dislodged from the edge of the continental shelf and rapidly swept into deeper waters. The valley could have been formed by the force generated by this powerful current of water and the newly displaced continental material. Because sand is found in the continental shelf, such an event could have transported the sand into the ocean's depths.

Some Dutch scientists had observed miniature, muddy-water avalanches in laboratory experiments. In these studies, the deposits laid down by the avalanches were always neatly graded with the finest grained material on the top and the coarsest on the bottom. The sand layers found in cores from the abyssal plain showed exactly the same gradations.

If avalanches had caused the sand to be transported to the abyssal plain, what had caused the avalanches? Earthquakes appeared to be the answer to that question. Ewing remembered hearing about the earthquake of 1929, which was located near the Grand Banks fishing area. The center of the earthquake was at the base of the continental shelf off Newfoundland. After the quake, underwater telegraphy cables—hundreds of miles apart—broke in a regular time sequence. The farther away from the quake's epicenter, the later the break time. Indeed, the last cable broke about 12 hours after the first cable was destroyed.

It appeared that a fast-moving, muddy-water avalanche had snapped the cables as it swept along the floor of the ocean.

These avalanches were called turbidity currents by the Dutch scientists. It became clear that such currents had been generated many times in the history of the Atlantic Ocean. The flat abyssal plain of the Atlantic could now be understood. After four years of study, the scientists determined that the plain was made up of two different sorts of materials. One category consisted of deposits of marine animal shells and other sediments that descended from the surface waters. The other was a mixture of sand and clay that was transported by occasional avalanches—turbidity currents—that originated at the edge of the continental shelf. The mystery of the misplaced sand had been solved.

There was more than a purely scientific interest in the study of the abyssal plain. It seemed possible that one source of petroleum is formed when concentrations of near-microscopic-size marine animals are overwhelmed by an underwater avalanche. If buried in an oxygen-free environment, the animals' carbon-based molecules are turned slowly into crude oil. Petroleum explorations are now under way in areas where deposits from turbidity currents have been located.

The Risk Factor

Ewing, unlike most research administrators, refused to be bound to a desk. Until he approached his 60th year, he continued to lead as many expeditions as possible. If he missed a departure because of his other commitments, he would join the expedition later.

In early 1954, on the third voyage of the research vessel *Vema*, he was able to sail with the ship. A few hundred miles off the mid-Atlantic coast on the morning of January 13, *Vema* was running in high seas and rolling and pitching violently. Ewing was on deck to check the *Vema*'s position when the ropes holding four oil drums gave way. The drums began to roll around the deck and smashed into some gear including precious research instruments. Ewing's brother John and the first and second mates came up to help Ewing resecure the drums. They had just retied them when a particularly large wave washed the drums and the people overboard.

Charles Wilkie, the first mate, was lost. The other three people were rescued. This was a near miracle because the *Vema* had sailed on for several minutes—and over a mile—before the accident was discovered. However, under the skilled hand of the captain, the *Vema* was turned around and brought back to the scene of the disaster. The second mate, Mike Brown, and John Ewing were not in bad shape when they were rescued. Brown actually stood his next watch. Maurice Ewing, the last person rescued, had a concussion and his left side was temporarily paralyzed. He was admitted to a hospital in Bermuda for recuperation. Nevertheless, he rejoined the expedition before *Vema* turned back to New York. No one knows whether Ewing thought of Wegener's fate during this misadventure. However, he must have been reminded that the high cost of field research might include the researcher's own life. Ewing walked with a limp ever after.

The Heezen Episode

One of the most colorful characters in Ewing's immediate circle of coworkers was Bruce Heezen. Heezen was born in Iowa in 1924. Like Ewing, he grew up far from the sea, and like Ewing, he became a great oceanographer. He also shared Ewing's dislike for authority and paperwork. Indeed, folklore has it that he refused to fill out the necessary papers to renew his driver's license. One imagines that Heezen saw Ewing as a kindred spirit. Together, at the Lamont observatory, they entered wholeheartedly into battles with bureaucrats and conditions at sea and in their attempts to persuade nature to give up its secrets.

Heezen was among the first to bring a woman into geophysical research. Before the 1950s, there were very few women working in the earth sciences. Women graduates in geology most often went into full-time teaching or went to work for industrial organizations such as major oil companies. In such industrial jobs, their primary function was analyzing rock samples.

Heezen had a career-long association with Marie Tharp that led to joint authorship of several important papers. Tharp was also responsible for both technical and theoretical contributions to the scientific program at Lamont. In fact, her success at Lamont was probably a major factor in breaking down the prevailing prejudices against women working in the earth sciences.

It was Tharp who first recognized that the Mid-Atlantic Ridge had a central depression that ran its full length. In other words, the so-called ridge had a major rift—or seam—down its center. Therefore, there were actually two ridges—somewhat like the left and right shoulders of a broad valley.

Early in the 1950s, Ewing and his team mapped the North Atlantic Ocean by using a seismic profiler. This instrument, developed in part by the team, produced seismographs—graphic recordings of sound waves—that showed the sea bottom and the layers of sediment and basaltic rock beneath it. Ewing's technique involved the release of a string of small blocks of TNT into the ship's wake. The TNT was timed to explode in a regular sequence. At the proper moment, a waterproof sound recorder, attached to the ship by a rope, was thrown over the stern and allowed to float in the water. The rope attached to the recorder was unreeled as the ship continued on its path. After the detonations went off and the echoes were recorded, the apparatus was reeled back aboard. The paper tape records were removed for immediate processing and the whole sequence was quickly repeated.

These continuous records were a marvel for their time. However, they were very difficult to interpret unless one had special training. The technique produced a graph with a highly exaggerated vertical dimension so that the profile of the bottom features appeared far more jagged than it really was.

A diagram produced by the profiler shows the area on either side of the Mid-Atlantic Ridge as a roughly symmetrical pattern of foothills and increasingly tall seamounts. However, because of the exaggerated, spiky profile of the seismograph, a scientist had to have a superior knowledge of underwater topography to identify a central rift valley between the ridges. After seeing only six profiles of the ridge—taken from slightly different latitudes—Tharp was able to identify that valley.

Bruce Heezen did not accept Tharp's discovery for a full year. Finally, he was convinced when presented with information from two different maps. On one map, Tharp had plotted an outline of the rift valley from her oceanographic information. This map was overlaid by a transparency showing a map of the same underwater area and plotted with the locations of earthquake activity. It was apparent—even to Heezen—that the centers of the earthquakes all fell in or near the rift valley.

Next, the scientists drew a map showing worldwide earthquake activity. When the earthquake centers were connected by lines, a pattern of boundaries seemed to emerge. No one at the time saw this configuration as plate boundaries. However, some scientists began to believe that a line of earthquake centers might indicate the presence of a ridge-rift structure. This possibility was confirmed by an investigation of the line of earthquake centers along the middle of the entire Atlantic Ocean.

Ridge-rift structures were found in other oceans, as well. In fact, a ridge-rift was found in the Indian Ocean exactly where Tharp predicted it would be found. Interesting and unexpected new features were discovered during these studies. In some areas, the underwater earthquake belt was found to coincide with very deep valleys rather than the ridge-rift type features of the Mid-Atlantic. Scientists also discovered that there is no ridge-rift in the center of the Pacific. Instead, they found only volcanic hotspots—active, underwater volcanoes—such as those that formed the Hawaiian Islands.

These peculiarities led to confusion and bickering among the experts. Later, Tharp's identification of earthquake activity at the ridge-rift sites would prove paramount in explaining the forces that moved the huge crustal plates. At first, the geologists, geophysicists, and oceanographers could make no sense of what they were observing. A few more years of study were needed before the picture would became clear. However, it was apparent that Ewing's team had triggered a process that could not be stopped.

Unfortunately, the partnership between Ewing and Heezen came apart before the new theory of plate tectonics was accepted. One of Heezen's problems as a scientist was his difficulty in controlling his enthusiasms. Rather early in his association with Ewing and the Lamont observatory, he became convinced that the best explanation of major geological changes was global expansion. He thought that the earth had become larger over geologic time. He stubbornly held to this idea in the face of

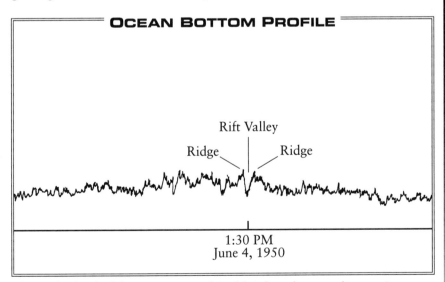

This is the kind of fuzzy image produced by the echo sounding equipment on Vema. The profile is much more jagged than the actual ocean floor because the horizontal dimension is shortened to save space on the paper strip. It was from such profiles that Marie Tharp was able to discern the structure of the Mid-Atlantic Ridge-Rift.

disbelief by his colleagues. One might sense that he adopted the unpopular idea of an expanding earth for the sheer joy of argument. A less generous view was that he used his debating skills as a way of getting attention.

Ewing thought that Heezen's idea of an expanding earth was absurd. In 1960, the two men stopped working on cooperative projects because of their disagreement. In 1962, Ewing's faith in Heezen was again shaken. While on duty as expedition leader, Heezen left his ship and was gone for three weeks. He did not tell anyone where he was going or what he planned to do. During the time that Ewing was the director of Lamont Geological Observatory laboratory. Heezen never again led an expedition.

Another Misunderstanding

The mid-1960s proved to be a crucial period in the revolution in the earth sciences. Some of the graduate students at Lamont were finishing the studies that would clinch victory for the plate tectonic theory—as will be described in Chapter 10.

In the late spring of 1966, before the younger scientists had fully completed their projects, a major international conference was held in Moscow. Heezen and a few other Lamont people were scheduled to attend this meeting. All the members of the Lamont delegation were strongly urged not to reveal what the graduate students had discovered. There were two good reasons for this suggestion. First, the young scientists had not yet completely reviewed and corrected their work. Second, it was understood that they would announce their own discoveries and get full credit for them.

Heezen could not resist the temptation to talk. He held a news conference that generated articles in the *New York Times* and *Time* magazine. Ewing was very unhappy.

Perhaps Heezen thought he was justified in speaking off-the-cuff in Moscow. He had been the first senior scientist at Lamont to understand the potential significance of observations made by the younger team members. Indeed, Heezen had encouraged further investigations as a way to broaden their initial discoveries. To his credit, he did not reveal exactly what had been discovered by the Lamont group. Heezen focused on information that had been developed by people from the U.S. Geological Survey. Mainly, he talked to the journalists about the long-term reversals in the earth's magnetic field. Perhaps his colleagues were most angered when Heezen spoke of the risk to human life. He told reporters that the earth could be bathed in cosmic rays when the next reversal took place. Clearly, such a conclusion excited much interest from the newspeople. However, there was little scientific basis for it. Furthermore, any such event was unlikely to occur until thousands of years had passed. Heezen's behavior looked more like showmanship than solid scientific thinking.

Heezen stayed on at Lamont through all these crises and, indeed, well after Ewing had retired. In fact, he led one expedition under Ewing's successor. Heezen was still affiliated with Columbia University when the U.S. Navy invited him to go on an expedition aboard a nuclear submarine. He planned to study the ocean floor just south of Iceland. While conducting this project, Heezen died of a heart attack on the first day of summer in 1977. Although he probably never believed in Wegener's theory of continental drift, Heezen was a worthy inheritor of the older man's legacy—he died while conducting his research. Wegener would have approved.

Ewing and the Theory

Over the years, the Lamont Geological Observatory became a leading center for the development of the theory of plate tectonics. However, Ewing, the director from 1949 to 1970, was not a strong advocate of this theory.

His true views on the theory may never be known. Although his findings contributed crucial evidence about plate tectonics,

Ewing did not see himself as the leader of a scientific revolution. He might be characterized as the stubborn skeptic of plate tectonics theory. Over the 20-year span from the late 1940s to the mid-1960s, he sometimes sounded like a true believer and sometimes like an opponent. He was an intense person, and this characteristic probably led him to say things that he later regretted. Although he finally accepted the theory in the fall of 1966, Ewing was one of the last important geophysicists to do so.

Ewing's personal life was fraught with problems such as the death of a young son and two divorces. His professional life, too, was difficult. Ewing's feud with Heezen must have been troubling. He had difficulties with university administrators, government bureaucrats, faulty equipment, and bad weather on his frequent oceanic expeditions. However, throughout his long career, Ewing was a productive and brilliant scientist. In addition to his role in the development of the theory of plate tectonics, Ewing was a major contributor to the scientific literature in general geology. Over a 50-year period, he authored or coauthored more than 150 important research articles. One of his most noteworthy contributions was a book published in 1957, *Elastic Waves in Layered Media*. W. S. Jardetzky, a mathematician, and Frank Press, later science adviser to President Lyndon Johnson and then director of the National Academy of Sciences, collaborated with Ewing on the book. It consists of highly technical information about the measurement of sound waves and their echoes. This book and many of his research reports are still in use. In addition, scientists continue to employ the research techniques that Ewing helped to develop.

By the time Ewing retired in 1970, the original name of his organization had been changed to the Lamont-Doherty Geological Observatory. During the time of his directorship—a 21-year period—his staff had increased from 20 people in 1949 to more than 400 members. His organization had collected more information about the ocean floor than any other laboratory in the world.

6

Some Magnetic Personalities

While Ewing was conducting his research at sea, a team of three young scientists were content to work on dry land. The work of Richard Doell, Allan Cox, and Brent Dalrymple—all graduates of the University of California at Berkeley—covered three exotic areas of research. They investigated the magnetism detected in rocks, techniques to determine the age of those rocks, and the reversals of the earth's magnetic poles.

When they began their research, Cox, Doell, and Dalrymple were little interested in controversial theories such as continental drift. The three scientists saw their studies as a means to advance the general knowledge of geological science. The team could not have foreseen that their work would someday confirm the findings of the Lamont scientists and help solve the puzzle of continental drift.

During their early years of study at Berkeley, no one anticipated that Cox, Doell, and Dalrymple would all become respected scientists. As young men, they fit the stereotype of laid-back California teenagers.

Doell, born in 1923, graduated from high school in 1940 with an interest in mathematics and some skill in building model airplanes. That fall, he entered the University of California at Los Angeles as a mathematics major. While in his first year at college, Doell's social life was a great success, but his academic record was not. His college days were temporarily halted when he was drafted into the army soon after the outbreak of World War II. During most of the war, Doell was stationed at the University of Oklahoma in Norman, where he took courses in civil engineering while completing officer training. After the war, he enrolled in the physics department at the University of California at Berkeley. Again, his academic work was unsatisfactory, and in 1947, he dropped out of school.

Doell drifted for a time before he was offered a job with an oil-prospecting firm. One of the senior geologists at the firm befriended him and tutored him in mineralogy. Over a two-and-a-half-year period, Doell learned the techniques of echo sounding—the use of sound waves to identify subsurface rock formations. He helped search for oil in the Santa Barbara Channel in California, Bahrain in the Middle East, and the Calgary area of western Canada. Doell found that he liked geology and decided to return to college. At that time, his main ambition was to gain advancement in the oil-prospecting firm and thought that some additional education in geology would help.

Doell's previous courses at UCLA and the University of Oklahoma allowed him to obtain advanced standing in the geophysics program at Berkeley. His academic performance greatly improved. Once again, Richard Doell was influenced by a senior scientist. With the help and encouragement of Professor John Verhoogen, Doell's work at Berkeley was successful. In 1952, after receiving his undergraduate degree, he enrolled in their doctoral program and received his doctorate in 1956.

By now, Doell's career ambitions were changing. He accepted a teaching job at the University of Toronto and then taught for one year at the Massachusetts Institute of Technology (MIT). In 1958, Doell returned to Berkeley to teach summer school between semesters at MIT. Allan Cox was enrolled at Berkeley, finishing his own doctoral program. This coincidence allowed the two men to resume a friendship that had begun in 1955. At that time, Doell was completing his graduate research in geophysics and Cox was a senior in the same department. During that summer, the young men became acquainted when they were both members of a field expedition team. After the expedition, Doell trained Cox to use some of the complicated equipment in one of the geophysics laboratories. When they met again three years later, the geophysicists found that they shared many common interests. Soon, a collaboration was initiated.

Allan Cox, two years younger than Dick Doell, was also a native Californian. However, his academic experiences were somewhat more conventional than Doell's. Cox was a voracious reader, and his teachers encouraged him to attend college. His parents were supportive if unenthusiastic about the matter. Cox attended UC Berkeley for the summer session of 1944, but dropped out before the fall term began. As an option to military service, he enlisted in the merchant marine. When released in 1948, Cox re-enrolled at Berkeley as a chemistry major. At that time, he had no particular goals, and his college grades were mediocre.

A fellow chemistry student at Berkeley introduced Cox to some friends at the U.S. Geological Survey. Soon, Cox was invited to take a summer job doing field work in Alaska. This job completely changed his attitude toward scientific work. Cox liked the people at the Geological Survey and the rigors of working outdoors. After returning to his chemistry classes in the fall, he began to think of switching his major to geology. In fact, Cox was so interested in geology that his chemistry grades became even worse. He lost his college deferment and was drafted into the army. From late 1951 until late 1953, he served in the Signal Corps. The time was not wasted, however, because Cox learned a great deal about sophisticated electronic equipment while attending a specialized training school.

In the meantime, Cox corresponded with his friends in the U.S. Geological Survey. His new interest in geology led him to re-enter Berkeley in 1954 and redirect his life toward a career in geophysics. While at Berkeley, Cox continued his involvement with the Geological Survey. On weekends during the school year, he volunteered to work for the Geological Survey, and in the summer, was given a paid job in Alaska. Cox was helped and encouraged during this period by Clyde Warhaftig, then a professional geologist with the Geological Survey and later a professor at Berkeley.

In 1955, after completing his undergraduate studies, Cox enrolled in Berkeley's doctoral program. Early in his graduate training, he was introduced to new responsibilities and ideas. He was soon placed in charge of the most sensitive laboratory equipment—a position once held by Doell. At about the same time, Cox became acquainted with the theory of continental drift through the teachings of Doell's friend and former mentor John Verhoogen. The theory was still highly controversial and unacceptable to most scientists. Later, Verhoogen's ideas on this subject would prove vitally important to Cox.

During the summer of 1958, Doell decided that he preferred to live and work in his native state of California. He wanted to pursue full-time research on rock magnetism and to continue his collaboration with Cox. However, he had a contractual obligation to complete his teaching duties at MIT.

After his return to MIT that fall, Doell discovered that he required surgery for skin cancer. His illness was another factor that motivated Doell to move back to California. Both he and his wife wanted to be closer to their families. Doell completed his commitment to MIT in the spring of 1959. He then returned to California and resumed his partnership with Cox.

In the spring of 1959, Cox finished his doctoral studies and accepted a job with his old friends at the U.S. Geological Survey at Menlo Park, California. After moving back to California, Doell also joined the survey and the two geophysicists continued their collaboration. They began the Rock Magnetic Project in an unused building that was little more than a tarpaper shack—so much for the glamour of the scientific laboratory.

Not long after their partnership resumed, Cox and Doell completed a survey of the scientific literature on rock magnetism. Their report, which was published in 1960, was well received by their fellow geologists. Early in their careers, therefore, the young scientists had become well known for their work on magnetism.

The two scientists then turned their attention from the study of scientific literature to field research. They collected large numbers of volcanic rock samples and tested them for the strength and orientation of their magnetism. Cox and Doell soon needed additional technical help and access to more advanced equipment. However, their resources were limited. They turned to a friend, Brent Dalrymple, who was then a graduate student at UC Berkeley. Cox and Doell hoped that Dalrymple would be able to allow them to use some of the laboratory facilities at the university. Luckily, this was possible.

When Dalrymple finished his own graduate studies in the summer of 1961, he joined Cox and Doell at Menlo Park. Dalrymple's academic background both paralleled and complemented that of his colleagues. Soon, the three launched a major research program to study the magnetic characteristics of rocks.

The Nature of Rock Magnetism

About 200 B.C., the Chinese discovered that when certain special rocks were attached to pieces of wood and floated in water, the rocks always turned toward the same location. If pushed into a new position, they returned to their original alignment. The Chinese soon realized that one end of the rock always pointed in a northerly direction, while the other end pointed to the south. This primitive compass was always dependable—no matter how far from home. Therefore, the traveler, whether on land or sea, knew the direction in which he or she was heading.

During the Middle Ages, European sailors began using a more advanced form of such compasses. The Europeans called the special rocks lodestones, but had no idea how they gained their power. Indeed, they knew little about magnetism or its properties. Years later, it became evident that lodestones contained a high concentration of iron—the metal easiest to magnetize.

In the early 1800s, while conducting basic research on the properties of electricity, some scientists recognized the relationship between electricity and magnetism. Gradually, the link became clearer and more explainable. For example, when an electric current flows through a wire, it generates a small magnetic field all around the wire. If one wraps this wire around a piece of iron, the electrical energy flowing through the wire will influence the arrangement of the iron atoms. Ordinarily, these atoms—each with its own small electrical charge—have no particular arrangement. However, the electricity in the wire will force some of the atoms to realign themselves at right angles to the flowing electrical current. When a large number of the atoms are aligned in the same direction, the piece of iron will be magnetized. Even after the wire is removed, the iron will remain magnetized and will attract other pieces of unmagnetized iron.

The new magnet has two locations where the magnetic force is strongest. These are called the magnetic poles. By tradition, one pole is designated the north pole and the other the south pole. Another magnet brought close to the new magnet may be attracted or repelled by it. The attraction or repulsion depends on which pole is presented to the new magnet. If the north pole of one magnet is presented to the north pole of the other, the two magnets will repel one another. If the south pole of one is presented to the north pole of another, they will attract each other.

The earth itself is a huge but not very strong magnet. It is our good fortune that the magnetic poles of the earth are located close to the planet's geographic poles. The geographic poles of the earth are the center of its rotation—the spinning motion of the earth. The geographic poles define the most northerly and the most southerly places on the planet. The magnetic poles are close to, but do not exactly coincide with the geographic poles. No one knows why the magnetic poles are offset in this way. However, even with the offset, the magnetic and geographic poles are close enough to be useful in navigation.

People now know that lodestone—the special rock used in a primitive compass—was probably formed from volcanic material with a high iron content. When the volcanic material, lava, was hot and in a molten state, it was not magnetized. As it cooled, however, it reached a specific temperature that made the iron atoms very sensitive to the earth's magnetic field. The atoms were then aligned in one direction and pointed to the earth's magnetic poles.

In more recent times, scientists have found that many kinds of rocks, even those with only a modest amount of iron, can be magnetic. These rocks show a faint magnetic field when tested by a highly sensitive device called a magnetometer. This faint magnetism was first detected many years ago. However, no one thought much about it until the 1950s. By then, scientists knew that rocks acquired their magnetic properties from the earth's magnetic field. They had also discovered that rocks do not lose this property unless they are reheated to a temperature near their melting point. Because of the stability of the magnetic orientation, each rock sample was an historic record of the orientation of Earth's magnetic field when that rock gained its magnetism. Therefore, with the proper equipment, scientists might be able to date the magnetic orientation of each magnetized rock and better understand the history of the earth.

Dating the Ages of Rocks

Traditionally, geologists determine the age of rocks by examining the microscopic creatures that are buried in the rock. If some rocks contain no remains, the geologist looks for evidence in the rocks that lay above and below the layer being studied. Usually, this procedure gives a rough but useful age estimate.

After the 1940s, some rocks—especially volcanic types —could be more accurately dated by a complicated process using radioactive decay. Rocks and other materials contain unstable radioactive elements, which over thousands of years gradually decay—or break down—into stable elements. For example, uranium—an unstable element—gradually gives off radiation and decays into lead. If an ancient rock contains even a trace of uranium, it will also contain lead. By determining the amount of lead and the amount of uranium in the rock, scientists can obtain a quite precise indication of the rock's age. Since not all rocks contain uranium, geologists designed similar procedures for more common elements, such as potassium. The newer techniques are applicable to all types of rocks, even sedimentary rocks.

Initial Research on Rock Magnetism

By the late 1950s, it was possible to date most rocks with fair accuracy. When using new techniques to determine both the age and magnetic orientation of rock samples, scientists discovered an interesting phenomenon. They found that very old rocks were oriented toward a place hundreds of miles from the present location of the north magnetic pole. Younger rocks pointed to a location slightly closer, and the youngest rocks pointed to the present location of the magnetic pole. Rocks ranging in age from the oldest to the youngest showed an orderly sequence of orientation from one point to the other. Using this information, scientists constructed a diagram, using a curving line to show the path of apparent movement of the north magnetic pole over many millions of years.

The apparent movement of the north magnetic pole was called polar wandering. The idea that the magnetic poles might gradually change position over a long period of time was fairly strange.

The first studies on this apparently orderly movement of the north magnetic pole were done in Europe. Keith Runcorn from Cambridge University in England was a leader in this area of research. During a visit to North America in 1958, he analyzed a sampling of rocks from the eastern United States. His young assistant was Neil Opdyke, a student at the Lamont Geological

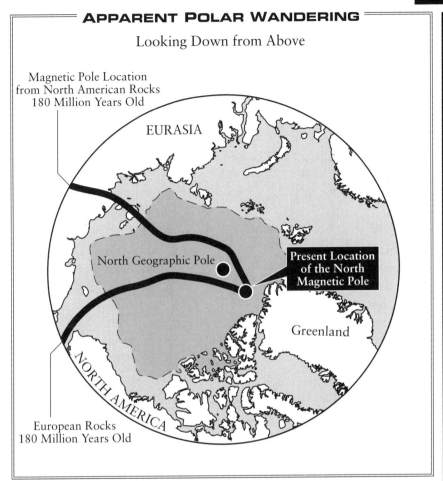

One path of apparent polar movement was generated by studies of the magnetism in rocks of various age found in Europe. The other path was based upon the magnetic orientation of rocks from North America. The two paths can be made to coincide if the shorelines of Europe and America are first joined and then gradually pulled apart.

Observatory in New York. Their investigation led to a comparison of the path of polar wandering generated by European rocks and the path generated by North American rocks.

At this point, the two scientists discovered something even stranger than polar wandering. When they diagramed the magnetic alignment of the oldest North American rocks, they observed that these rocks pointed to a different location than did the European rocks of the same age. As younger and younger rocks were tested, the diagrams indicated that the alignment of North American rocks gradually came closer to those of the European rocks. Finally, the youngest rocks from both areas pointed to the present location of the north magnetic pole. This seemed to mean that the magnetic orientation of each set of rocks began at a different place and gradually migrated to the same location.

It is possible that the two different paths were caused by the existence of two north magnetic poles, which converged in the recent past? This was hard to believe. When scientists examined the outcome of Runcorn's research, one explanation for this phenomenon seemed plausible, although startling. Perhaps the continents had once been very close, but had slowly drifted apart and rotated from their original locations. This possibility would explain the differences in the magnetic orientation of the two sets of rocks.

As might be expected, Runcorn's interpretation of this research was challenged. Many established scientists—particularly the Americans—hesitated before accepting Runcorn's new information on polar wandering as further evidence of continental drift. In spite of this opposition, Runcorn became convinced—contrary to his own previous beliefs—in the soundness of continental drift theory. Over the next few years, he encouraged Opdyke and some of his other colleagues to accept this point of view.

Polar Reversals

In the late 1950s, research on polar wandering continued while scientists from many countries—Japan, Australia, France, Iceland, the Netherlands, and Russia—studied geomagnetism. They investigated many areas, including the geographic distribution of magnetic rocks. In the course of this exploratory research, scientists found another strange fact. The magnetic orientation in some rocks from the Northern Hemisphere was completely opposite from what they had expected. The magnetic alignment of these rocks was toward the south magnetic pole rather than the north magnetic pole. According to the magnetic orientation detected by a magnetometer, the rocks seemed to register that the north magnetic pole had once been near the south geographic pole and vice-versa.

For most scientists, this was impossible to believe. They sought to explain this discovery by showing that a rock could change its magnetic orientation due to some natural event. Scientists hoped to attribute this change to some odd, internal modification or to an external force such as lightning. Otherwise, the scientific community had to admit that the entire magnetic field of the earth had, at times, switched locations.

The phenomenon of a rock having a southern rather than a northern polar orientation is called a polar reversal. In searching for an explanation of the mystery, some scientists even duplicated reversals in the laboratory by using complicated procedures that involved heating and cooling the rocks. However, these scientists could not explain how the artificially generated reversals could possibly occur in the natural world.

In fact, no one understood how the reversals took place. Nevertheless, the phenomenon had to be accepted in the face of many independent observations. It seemed that over a span of a few million years, the north and south magnetic poles actually had switched locations. Such polar reversals appear to have occurred many times during the history of the earth.

Once it was firmly established that such a thing as magnetic reversal could take place, the next challenge was to identify the precise times at which these changes occurred. The first research along such lines was done by an accomplished German geologist, Martin Rutten. In 1959, Rutten attempted to date magnetic reversals by studying lava from some Italian volcanoes. The German geologist gave up the task after a few studies had been completed. However, Rutten's investigations inspired Doell and Cox to continue his line of thought. Their research—begun in 1960—resulted in the development of a precise time scale covering the most recent 10 million years. The time scale would soon be seen to contain a key to the puzzle of continental drift.

Throughout the early years at Menlo Park, the research done by Cox's team focused on rocks produced by volcanic eruptions. Their first attempt at dating polar reversals showed that the normal—present-day—polarity went back almost 1 million years. It also showed that reverse polarity existed in the previous million-year time span. Before the 2-million-year mark, the polarity had been normal. Their report on this research was published in mid-1963. By that time, Dalrymple, their younger colleague at UC Berkeley, was a full-time member of the team.

The Australians Enter the Game

As Cox, Doell, and Dalrymple continued their studies, a similar effort commenced in Australia. This parallel research was done at the Australian National University at Canberra. Over the years, Canberra had developed a sister relationship with UC Berkeley in the areas of geology and geophysics. Students and faculty from the two universities moved back and forth with the help of various government-sponsored fellowships. Two of these exchanges were especially important.

In 1960, Jack Evernden, a senior researcher and faculty member at Berkeley, was invited to Canberra for a six-month visit. The department of geophysics at Canberra had just been reorganized, and the faculty wanted Evernden to help them set up a rock-dating laboratory. By duplicating Berkeley's equipment, he finished the facility in six months.

On the other side of the Pacific, an Australian, Ian McDougal, arrived at Berkeley for a year of study on a post-doctoral fellowship. His specific mission was to learn as much as possible about the rock-dating method used in the Berkeley laboratory.

When McDougal returned to Australia in 1962, he was put in charge of dating rocks. However, he was uneasy about using the temperamental equipment that Evernden had set up. McDougal did not want to disappoint his sponsors by producing poor results and made little headway with rock dating until 1963. That year, he read the first published results of the polarity reversal study by Cox and his team. Then McDougal went to work. For the next few years, there was some rivalry between the Australians and the Americans. It seemed that every time the Americans published a report on dating polar reversals, the Australians countered with a report of their research on the same time span. Since both groups were obtaining similar results, both became more confident of their work. The friendly competition was constructive, and the quality of the time scales gradually improved.

Refining the Time Scale

At first, Cox, Doell, and Dalrymple focused on the polarity record of the last 4 million years. They found it was relatively

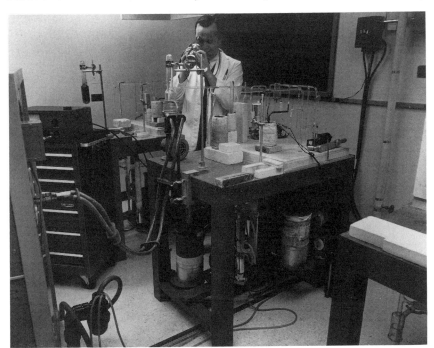

A Laboratory for Dating Rocks: The age of small samples of rock can be found by heating the rock and capturing the gasses given off. The amount of argon gas, when compared to the amount of potassium in the rock, reveals its age. (Courtesy of the U.S. Geological Survey)

easy to locate young lava deposits in that age range on or near the surface. It was a lucky accident that these fairly recent events contained the most important facts for studying continental drift.

Actually, the time period between 700,000 and 2,300,000 years ago held the crucial information. Within that period of reversed polarity, there were two brief segments of normal polarity. The research team gave special names to these short, normal intervals. The older and longer—about 200,000 years in duration—was called the Olduvai Event. The sample of volcanic, crystalline rock that showed this event had been taken from Olduvai Gorge in Africa, a major site for the study of humanlike fossils and stone tools. The newer and shorter—less than 100,000 years in duration—was called the Jaramillo Event. The rocks that revealed this event were lava samples found in New Mexico near a small stream of that name.

By early in 1966, the team of Cox, Doell, and Dalrymple had revised and refined their dating scale six times. Their competitors in Australia and elsewhere had devised four alternative versions. Each published time scale was backed by more evidence from rock samples and was a bit more accurate and more convincing than its predecessor.

These time scales are usually represented graphically by a series of columns. The length of each column represents a specific period of time. Intervals of normal magnetism are shown as black spaces in the column, and intervals of reverse magnetism are shown as white spaces.

Cox's graphic representation of polar reversals shows the brief intervals of normal polarity—the Olduvai and Jaramillo events—as distinctive black markers within an extended period of reverse polarity. Scientists involved in similar rock research could easily recognize these events. Indeed, physical oceanographers identified, although by different names, the same events when studying the magnetism of marine cores. For a while, however, very few scientists were aware of this parallel record. Even fewer understood the vital connection between the information supplied by land geologists on one hand and marine geologists on the other.

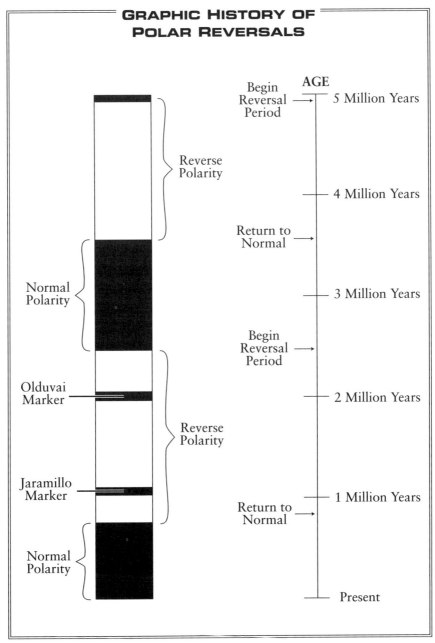

This diagram shows the time sequence of changes in the earth's magnetic field. The pattern was first established by using samples of volcanic lava that had solidified on the continental crust.

In a research report published in the early spring of 1966, Cox, Doell, and Dalrymple expressed the hope that their refined dating scale might be useful for "geologic correlations or other purposes." Their hope was soon realized. By using the new dating scale, land and marine geologists could better compare the magnetic reversals recorded in their own samples. In a matter of weeks after the 1966 paper appeared in print, geologists from both special fields recognized that rocks from the ocean floor and continental rocks revealed the same patterns of polarity—including the markers of the Olduvai and the Jaramillo events.

Aftermath

By 1972, the pattern of reversals had been traced back for 162 million years. The longest period of normal magnetism lasted almost 30 million years and ended about 84 million years ago. The records show 85 periods of normal polarity and 84 periods of reversed polarity. Over the years, scientists have extended the history of reversals. By 1996, the period extends back—with some gaps—for about 340 million years! However, the two most revealing occurrences, the Olduvai and the Jaramillo events, took place in comparatively modern times—less than 2.5 million years ago.

7
Moving Along to La Jolla

Well before Cox and his team at Menlo Park had begun studying the magnetism of continental rocks, scientists from the Scripps Institution of Oceanography at La Jolla, California, had been conducting similar studies on marine rocks. Although both research organizations are linked to the University of California, communication between the two was sometimes unsatisfactory. Possibly this happened because the two groups of scientists were involved in different areas of geology. Geophysicists were working at Menlo Park, and physical oceanographers at La Jolla. For whatever reason, the scientists felt little incentive to share information.

Indeed, after the mid-1950s, both groups were studying rock magnetism at exactly the same time. When the geophysicists analyzed continental rocks, the magnetometer revealed "reversals" to them. When the La Jolla group tested rocks from the seafloor, they saw "anomalies." At the time, no one realized that they were both seeing the same things, but calling them by different names. After the mid-1960s, the scientists belatedly realized that the two sets of magnetism studies had offered the opportunity to cross-check their observations. The scientists at La Jolla and Menlo Park had unwittingly delayed a major scientific breakthrough by several years.

Administration of the Institution

Scripps was headed by Roger Revelle from the late 1940s until 1964. Revelle was already a major figure in the earth sciences when he was made director. Because of his sound leadership, he later become a true statesman of science—especially in the area of environmental policy making.

Scripps had been an active research center prior to the outbreak of World War II in 1939. During the early 1930s, Roger Revelle was one of the young staff oceanographers. Throughout the war years, he served as a naval officer. In 1945, he was the chief oceanographer for the atom bomb tests on Bikini Atoll in the Pacific. A year later, Revelle became head of the Geophysics Branch at the newly formed Office of Naval Research in Washington, D.C. That same year, he recommended government support for Ewing's oceanographic research at Columbia University.

In 1948, the director of Scripps was about to retire. The older man asked Revelle to suggest a strategy for Scripps's future development. Revelle's eight years of government service had given him a good grasp of organizational matters, and he designed a workable plan. He proposed that Scripps be merged with a younger organization, the Marine Physical Laboratory at the University of California at San Diego. The proposal was accepted. Professor Carl Eckart, then head of the university's laboratory, was named director of the combined organization. Revelle was appointed deputy director. In 1951, Eckart retired and Revelle became the director.

The resources of U.C. San Diego and Scripps complemented one another. The university was able to offer a stable base for the new facility and graduate students to fill the many research positions. For its part, Scripps provided considerable prestige and an excellent endowment.

In 1947, while still deputy director, Revelle led Scripps's first major postwar expedition into the Pacific. The purpose of this expedition—named MIDPAC—was similar to the first Atlantic expedition organized by Lamont observatory in New York. The primary mission of both groups was to map the ocean floor by the use of echo sounding.

The Scripps's scientists found many surprises. First, the Pacific Ocean floor off the coast of California is consistently hilly. The floor of the Pacific is unlike the flat, abyssal plain that Ewing found in the Atlantic off the coast of New Jersey. Scientists had reasoned that the same condition would be found in the Pacific. Revelle's expedition proved otherwise.

The next surprises were discovered after the first seismic shots had been completed. These undersea explosions were detonated about 300 miles (480 km) offshore La Jolla. The results indicated that the sediment layer was about 900 feet (270 m) deep—about the length of three football fields—not the 3 *miles* (4.8 km) of sediment that scientists had previously believed. In addition, the seismic studies showed that the crust beneath the sediment was about 5 miles (8 km) thick, much thinner than the expected 18 miles (28.8 km). Before these studies, scientists had assumed that the seafloor of the Pacific was very ancient and would, therefore, have deep layers of sediment and a thick crust. Since Ewing's people found the same relatively thin crust in the Atlantic, it now seemed likely that all oceanic crust was rather thin.

Another surprise was the realization that heat was rising from the bottom of the ocean. The water near the seafloor is frigid, and even the faint heat that rises from below the crust can be measured by thermometers anchored on the bottom. More heat was being released than anyone had expected. This observation suggested that molten magma might be flowing beneath the ocean crust.

During Revelle's early years at Scripps, he invited a number of senior geologists to serve as guest scientists on the Scripps expeditions. These scientists, from the Sea-Floor Studies Section of the Navy Electronics Laboratory in San Diego, included Robert Dietz and his assistant, William Menard. Both sailed on the first MIDPAC expedition and collaborated on the analysis of the findings.

Later, Menard transferred to Scripps. In 1951, after Revelle took over as director, Menard became an expedition leader. Dietz joined the faculty at Princeton University. His theories on physical oceanography contributed to the major scientific breakthrough in 1966.

The Great Fractures

During the late 1940s, scientists at Scripps made their most important contributions to the concepts of continental drift and seafloor spreading. During their oceanographic research, they discovered two unusual features on the seafloor of the Pacific Ocean. Echo sounding located the so-called fracture zones, and later, magnetometers detected a pattern of magnetic stripes.

Russian oceanographers had found some indication of fracture zones in the Atlantic Ocean just after World War II. However, they had shown little interest in their find. After William Menard made his initial observation of fracture zones, these features became the focus of his work.

Menard's introduction to his lifelong study occurred quite by chance. In 1949, while working at the naval facility at San Diego, Menard accidentally missed the departure of a research vessel. In order to keep busy, he decided to look over some echo-sounding reports that had been around for years. In Menard's own words, these records were a "mess." He spent his time correcting obvious mathematical mistakes and drawing contour maps of the Pacific floor.

Menard discovered that years earlier an oceanographer from the U.S. Coast and Geodetic Survey had located a large crack in the floor of the Pacific off the town of Mendocino, California. The crack extended westward from the coast for hundreds of miles. Strangely, the profile of hills and valleys on the north side of the crack did not quite match up with the very similar profile on the south side. It looked as if one side of the crack had been pushed from its original location. This was very curious.

In 1950, nothing was known about the newly discovered fracture zones. Twenty years later, scientists determined that these faults appear at right angles to every ridge-rift structure and are located about 70 or so miles (112 km) apart. They realized that the cracks are caused, in part, by the movement of the great crustal plates. They are caused, too, by the differing rates of seafloor spreading that occur at various locations along the ridge-rift structure.

Menard set out to map all the cracks—fracture zones—in the Pacific seafloor. By 1952, he had located two more cracks south of the Mendocino fault. He named these the Murray Escarpment and the Maury Trough in honor of two noted geologists. Menard also discovered the Marquess Fracture—named after a nearby island chain—during a 1952 expedition south of the equator. It was not until 1956 that Bruce Heezen from Lamont found fracture zones in the South Atlantic, ranging east and west from the Mid-Atlantic Ridge.

Although William Menard, the most energetic investigator of fracture zones, never attained a directorship as Maurice Ewing had done, their careers were similar in many aspects. Both men loved expeditions and had many seagoing adventures. In 1964, Menard was on board the research vessel *Baird* south of Easter Island. He was on deck after a storm when a heavily encased, deep-sea camera came loose from its restraining ropes. It slid across the deck and hit Menard in the back. He was knocked unconscious and narrowly escaped being swept overboard. The injury caused three vertebrae to fuse together. Menard—like Ewing, a decade earlier—was impaired for the rest of his life.

Ewing and Menard spent their lives collecting information. Indeed, Menard admitted that he was addicted to research. Too, both men were opinionated in their scientific views and reluctant to accept the theory of continental drift. Menard claimed that he gave little serious consideration to the theory until late in 1966. Menard, unlike his colleague Heezen, did not die at sea. In 1986, he died in a hospital bed from cancer. He had been planning his next seagoing expedition.

More Magnetics

Research at the Scripps Institution went well during the early 1950s. New information about fracture zones allowed a more complete understanding of Menard's discovery. Interesting and important undersea features were found off the Pacific coast near Scripps. By coincidence, these new finds were located near areas of military interest. By using the same chartered research vessels—thereby saving much money—scientists attached to the navy could conduct defense-oriented studies and those from Scripps could carry on their oceanographic investigations. Much information was gained on these joint ventures. It is unfortunate that faulty interpretations of some underwater finds marred the record of this undertaking.

One such episode took place in 1954, when Ronald Mason, a British geophysicist from the Imperial College of London, was conducting research on magnetism found in seafloor sediment. Mason and a team of Scripps people—including Arthur Raff, a new team member—sailed on an expedition toward the west coast of Mexico. The scientists reasoned that the red clay found on the seafloor in that area probably contained iron and would show magnetic properties. Samples of the ancient clay, composed of countless layers of compressed sediment, were carefully collected over a 100-mile sq (260 km) area.

Thirty-foot (9 m) plugs, or cores, of clay were removed from the ocean floor at 20- (32 km) to 30-mile (48 km) intervals. Samples from the cores were labeled and taken to London by Mason. These samples, cut from spots along the full length of each plug, represented a time span of 10 million years. When tested with a sensitive magnetometer, several magnetic reversals were discovered. Although the pattern of reversals differed slightly from core to core, the similarities between the cores were detectable. Mason and Raff were convinced that they had discovered a record of polar reversals in the layers of sediment that made up the red clay. Before their find, rock magnetism reversals had been detected only in lava.

When Mason and Raff related their interpretation of the discovery, their colleagues scoffed at the idea. The two scientists were advised that they would appear foolish if they published an article about these findings. They reluctantly abandoned that line of research. Later, the scientific community realized the validity of their initial interpretation.

The next error was more serious. In 1954, the same year as his research on the red clay cores, Mason was invited to conduct a magnetic survey *across* the floor of the ocean south of San Diego. He hoped to gain magnetic information by towing a magnetometer that had been placed in a waterproof container. The instrument was pulled along by the ship. Mason had successfully carried out a similar survey in 1952, but on this voyage the magnetometer failed to work properly. Unfortunately, Mason and his colleague Raff did not obtain any information from the southward cruise of the ship.

The research vessel returned to San Diego before beginning the second part of the voyage, a northerly cruise toward Vancouver Island in Canada. The two scientists quickly repaired their defective equipment and made ready for sailing. The research vessel carefully followed a prescribed route. It sailed for hundreds of miles due west, turned north for a short distance, and then cruised hundreds of miles due east. This pattern was repeated until the ship had reached the northern limit of the expedition.

This survey was successful. The entire underwater area had been plotted, and Mason and Raff had collected a continuous stream of information.

If anything, there was too much information. Mason had developed a complicated mathematical procedure to analyze all the findings. At that time, there were no powerful computers, and Mason did the lengthy calculations by hand. It took a very long time.

Revelle was annoyed by the slow progress and feared that their government sponsors would think that Scripps was unproductive. Raff continued to collect more and more data, and Mason got further and further behind in his calculations. Finally,

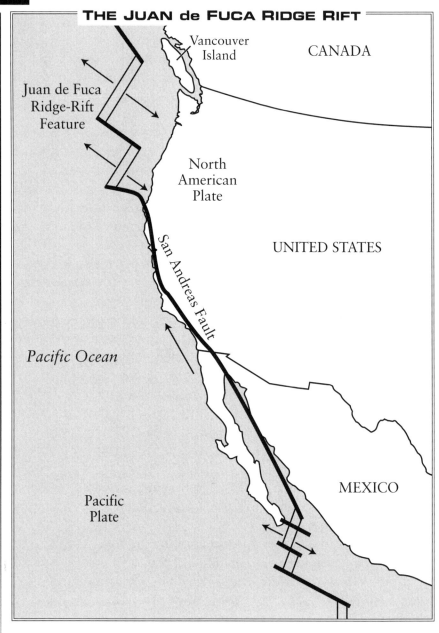

This map shows the general arrangement of the ridge-rift and connecting fractures west of Vancouver Island. This was the area of the ocean floor surveyed by Mason and Raff using their underwater magnetometer. (Courtesy of the U.S. Geological Survey)

Revelle asked the two scientists to avoid any new projects until they had written up their observations. The two scientists did not comply with Revelle's request.

Impatient with continued delays, Revelle instructed Menard and a newly hired staff member, Victor Vacquier, to analyze and interpret some of Mason's and Raff's earlier findings. Revelle hoped that this drastic move would cause the reluctant authors to prepare their work for publication. Revelle's strategy succeeded. In 1961, *two* reports by Mason and Raff appeared in print at exactly the same time as the paper by Menard and Vacquier. All parties, including the navy, were enthusiastic about conducting further studies of seafloor magnetism.

An analysis of the material collected by Mason and Raff uncovered startling information. Near the U.S. border with Canada—the northern limits of the voyage—the magnetometer had detected a strange pattern of magnetic fields on the ocean floor. Zones of relatively strong magnetism alternated with zones of relatively weak magnetism. When these zones are illustrated in graphic form by using a dark color to represent strong magnetism and a light color to represent weak magnetism, the pattern resembles zebra stripes. These magnetic zones or stripes are parallel to the Juan de Fuca Ridge-Rift.

Initially, these striped areas were a complete mystery. After a great deal of study, scientists came to understand the significance of the stripes. The areas of low magnetic intensity are places where the ocean crust registers one type of polarity—either reversed or normal—and the overlaying silt registers the reverse type. The opposite polarity of the layer of silt weakens the apparent strength of the magnetic field of the crust.

The areas of high magnetic intensity are those in which both the crust and the overlying silt have the same polarity. In such cases, the identical magnetic orientation of both layers strengthens the intensity of the magnetic field.

As some scientists sought to understand the significance of the magnetic stripes, others worked on a related mystery. The magnetometer showed that the length of each stripelike area was intermittently broken by large cracks on the ocean floor. On

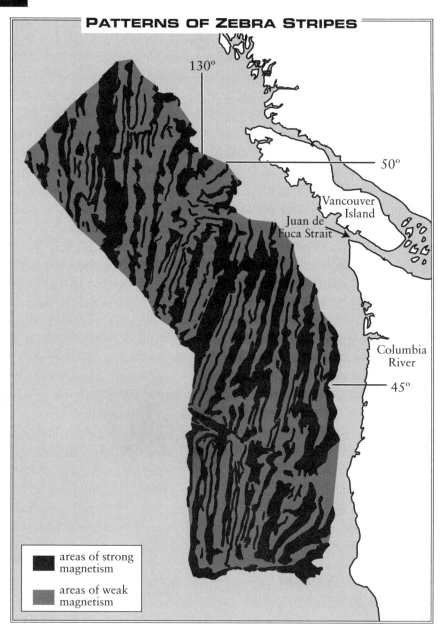

When the magnetometer readings from the Juan de Fuca region were analyzed, they showed a pattern of parallel stripes of stronger and weaker magnetism. Because of the jumble of seamounts in the area, the stripes have uneven edges.

either side of the fault, the continuation of the stripe was relocated either to the left or right of its previous path. Scientists now know that these cracks are, in reality, fracture zones similar to the larger cracks that Menard was then studying.

In the early 1960s, Menard and Vacquier had studied this type of misalignment using the hills and valleys of the ocean floor as markers. When they studied the conformation of the magnetic stripes, they concluded that the segments of each magnetic stripe had been in perfect alignment at some time in the past. By using diagrams of stripes, Menard and Vacquier were able to line up the pattern on one side of the fault so that it matched with its twin on the other side. The scientists realized that some enormous force had rearranged the floor of the ocean. In some cases, this movement was extensive. In fact, the north and south sides of the Mendocino fault were out of alignment by 640 miles (1,024 km). While this discovery was impressive, it was not well understood.

In addition to this mystery, no one grasped the complete significance of the magnetic pattern of zebralike stripes. The fact that a very similar pattern of stripes was recorded on either side of the Juan de Fuca Ridge was totally overlooked. The research team—Menard, Vacquier, Raff, Mason, and others—were so focused on the problem of the mismatched stripes that they missed seeing the symmetry of the pattern. Indeed, the symmetry was difficult to discern because of the hilly nature of the seafloor. Later, Menard explained that everyone's eyes were tired from poring over complicated maps and diagrams. It was not until 1966 that Menard, with the help of his research assistant, Tanya Atwater, comprehended the full meaning of what they had seen but not understood in 1961.

8

The Geopoets

William Menard, a compulsive observer, thought that most theory making was a waste of time. Nevertheless, he did help many scientists who developed theories. Six of these scientists are particularly important as successors to Alfred Wegener.

As a term of disrespect, Wegener was once called a *geopoet*—a geologist who constructs imaginative and complicated theories about the earth. Gradually, the term gained a positive rather than a negative meaning. Like Wegener, these six scientists are known as *geopoets*.

One of these geopoets was the physical oceanographer Robert Dietz. Dietz had received his doctorate from the University of Illinois shortly before the beginning of World War II. When the war started, Dietz activated his reserve commission in the army. Since he was already a licensed pilot, he was taken into the U.S. Army Air Corps. In 1945, Dietz was assigned to take command of a squadron of twin-engine bombers. However, the war ended before he could take up his posting.

Throughout his postwar career, Dietz wrote imaginative articles about such topics as the impact of meteorites on the earth and the origin of the moon's craters. Sometimes his ideas were criticized by more conservative geophysicists.

After the war, Robert Dietz was William Menard's immediate supervisor at the Naval Electronic Laboratory in San Diego. In 1947, the two men sailed on the first Scripps MIDPAC expedition. Unlike Menard, who sailed on many such research cruises, MIDPAC was the only Scripps expedition for Dietz. He soon realized that his interests lay in interpreting information rather than gathering information.

Dietz was granted a Fulbright Fellowship in the early 1950s and spent the time teaching geology at a Japanese university. In 1954, he was transferred to the London branch of the U.S. Office of Naval Research. In London, Dietz summarized the research of European geologists as a way to inform American scientists of current scholarship. He remained there until 1958 and then returned to the Naval Electronics Laboratory in San Diego. While working for the navy, Dietz participated in several record-breaking deep-sea explorations in the Mediterranean and the far Pacific.

Harry Hess Favors Seafloor Spreading

Harry Hess was another geophysicist whose ideas helped forge the theory of plate tectonics. Hess did his graduate studies at Princeton in the early 1930s, where he came under the influence of Professor Richard Field. In the late 1930s, Field persuaded Hess, Maurice Ewing, and others to begin research on underwater geology.

During World War II, Hess served in the navy and remained in the service as a naval reservist after the war. During those postwar years, he completed several special assignments for the navy and ultimately retired with the rank of rear admiral.

While on active duty during the war, he discovered 160 submerged, extinct volcanoes in the deep Pacific. Hess called them "guyots" in honor of Arnold Henry Guyot, an outstanding, Swiss/American geologist of the 1800s.

Hess saw the guyots as ancient Pacific island volcanoes that had been slowly eroded until they disappeared beneath the waves. Hess thought that most of these submerged mountains were about 500 million years old. He also reasoned that research on these ancient, now flat-topped, seamounts would provide a better understanding of the evolution of the seafloor.

After underwater dredges dug deeply into the sides and tops of the guyots, the sediment revealed that the age of the volcanoes was about 180 million years old or less. This was far younger than the 500 million years that Hess had advanced. He had already published papers in which his estimate of the greater age had been a key ingredient. Indeed, he had even developed a theoretical explanation for his estimate.

Sometimes, when some new observations contradict a well-reasoned estimate or theory, the theoretician says little and hopes that the discredited idea will soon be forgotten. Other theorists develop a replacement theory—perhaps one that is even more difficult to accept. Hess followed the second course. He was not dismayed when his dating proved inaccurate. Using some of his old ideas, he set out to develop a new theory that would account for the relatively young age of his guyots.

Dietz and Hess Share Some Ideas

In 1961, Robert Dietz stated that molten magma rises up along the center line of a ridge-rift structure—the boundary between two oceanic plates. According to his theory, the flow of the magma drags the overlying crust away from the central rift, and the edges of the plates slowly separate. Molten magma that has oozed up through the rift then fills in the space between the slowly separating plates.

Dietz also proposed that the old crust is ultimately recycled when it returns to the asthenoshere—the layer of molten lava

He believed that the deep-sea trenches indicate where the ocean crust follows the descending contour of the continental crust and eventually sinks back into the asthenosphere. Lastly, he said that Menard's fracture zones indicate the boundaries between fast- and slow-moving segments of the earth's crust. Although this idea later proved to be only partially correct, it seemed a reasonable theory at the time.

Dietz showed a draft of his manuscript to William Menard, his friend and colleague. Menard was astonished by what he read. Just a short time before, he had examined the draft of a manuscript written by Harry Hess. Amazingly, the two papers contained many of the same ideas and even some similar wording.

Menard knew that Dietz and Hess had talked to one another about new developments in geophysics. However, he was positive that the manuscripts were written independently. In the history of science, such coincidences have occurred many times.

Although there were many similarities in the manuscripts, there were differences, as well. Hess began the draft of his reworked theory with the dramatic birth of the earth's solar system. Dietz began his paper with a more prosaic discussion of the importance of ridge-rifts in the evolution of the ocean floor.

Hess had developed a comprehensive and involved theory so that he could explain the age of his guyots—the submerged, dormant volcanoes. Hess's new theory began with a mass of cosmic dust and gasses, which gradually formed the sun and all the planets. He explained that the newly born earth was a molten, radioactive ball. From within the new planet, the lighter, less dense, molten rock flowed upward to the surface and cooled to form thick "islands" in a sea of molten magma. This thick but relatively lightweight crust—composed of granite rock—formed the continents. Later in geological time, a heavier, denser, molten rock—mainly basalt—rose to the surface. The newer material cooled and solidified into a relatively thin, but weighty, skin that became the ocean crust.

Hess continued his story with the birth of the guyots. At various times in Earth's history, a rising plume of molten magma rose up to form a "hot spot" under the ocean crust. Sometimes,

The Assal Rift in Africa: The fault line comes out of the Gulf of Aden. This is one of the few places where volcanic guyots can be seen above the surface of the sea. (Courtesy of the National Geographic Society)

the magma broke through a weakened spot in the crust and a single, active volcano, or guyot, was formed. Over time, the ocean crust moved away from the hot spot, carrying the volcano with it. Soon, another weak place in the crust would move over the hot spot, and a new active volcano would be born.

As the older, dying volcano continued to be moved away, it became dormant, then eroded, and finally disappeared below the waves, where it continued to erode. All the while, the process of seafloor spreading carried the guyot toward a deep-sea trench where ocean crust and continental crust collide. The ocean crust, carrying the guyot, was pushed down along the edge of the thick, continental crust. Sometimes the guyot and other material of the oceanic crust was scraped off and became part of the continental mass. However, most of the crust, with the guyot attached, sank into the asthenosphere and became, once again, molten rock. To Hess, this theory explained why there were no elderly guyots. They had all been recycled.

THE EVOLUTION OF A GUYOT

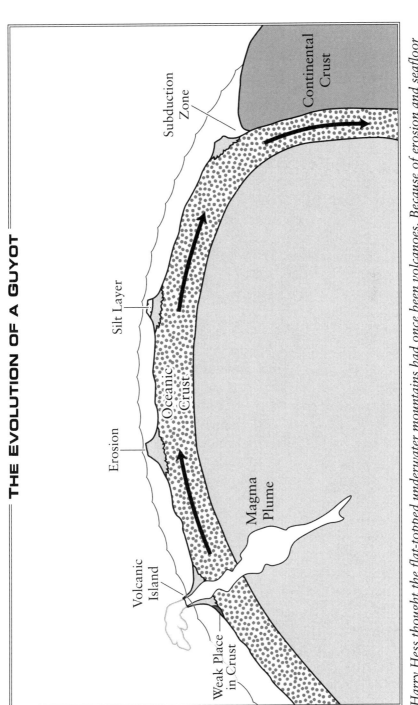

Harry Hess thought the flat-topped underwater mountains had once been volcanoes. Because of erosion and seafloor spreading, they submerged. Eventually, they are recycled along with the ocean crust upon which they ride.

The theories of both Dietz and Hess had been based on the same novel and disputed ideas—seafloor spreading and the recycling of old crust. Many of their conclusions about the ideas were almost identical. Hess's ideas, however, were more imaginative and far-reaching as he told the life story of guyots.

Hess might have been considered a crank or a fool by traditional geologists. He was, however, a respected member of the scientific establishment. Also, he disarmed his critics by calling his story geopoetry. His work attracted attention, but not always approval.

Menard could not decide whether Dietz or Hess had been the first to develop a comprehensive theory about seafloor spreading. The scientific community always places great importance on the order in which new information is published. Indeed, the first author to publish an article or book about a new theory or research project gains status within his or her field.

William Menard was never able to determine whether Robert Dietz or Harry Hess developed the first major step toward a theory of seafloor spreading. Happily, no dispute developed about the primacy of the manuscripts. Each author gave credit to the other!

Robert Dietz published his paper slightly before Harry Hess's publication of 1962. Dietz continued to work for the government as a staff scientist with the U.S. Coast and Geodetic Survey in Washington, D.C. Dietz's last and most satisfying job was as a professor of geology at Arizona State University at Tempe.

After the publication of his paper, Hess also went on to other matters. He collaborated on a project to drill a hole through the oceanic crust. Unfortunately, that project was a failure. Hess completed his scientific career by rejoining the faculty at Princeton.

In 1969, Harry Hess died while working at Woods Hole Oceanographic Institution in Massachusetts. At the time, he was chairman of a meeting of the Space Science Board. The scientists were working on a revision of the research agenda for lunar exploration. During his very last meeting, Hess was, as always, involved with the cutting edge of science.

The basic truth of Hess's theory of seamount evolution was later confirmed by a detailed study of the chain of islands that include the six inhabited islands that make up the state of Hawaii. The entire chain of islands includes the Hawaiian islands, small uninhabited islands, guyots, and Midway Island—a coral atoll built on the base of an extinct volcano. The chain covers a distance of about 1,600 miles (2,560 km).

It is now believed that the chain was built in a sequence—one island at a time—during a period of more than 100 million years. Each island was born when the Pacific Ocean plate moved over a very long-lived hot spot, which lies about 2,000 miles (3,200 km) southwest of California. Much as Hess had predicted, the new volcanic island was then carried away by the slowly moving oceanic plate. A newer volcano was formed over the hot spot, and the process was repeated.

Indeed, Hess's theory of seafloor spreading was strengthened by dating the rock found on the islands of the Hawaiian chain. The age of the islands was found to decrease from the most western to the most eastern island. Lava from Kanai, at the western end of the main Hawaiian cluster, is 12 million years old. The lava on the island of Hawaii—the youngest, largest, and most easterly island of the Hawaiian chain—measures only 1 million years old. This island has Mount Kilauea, a spectacular, active volcano. Because of Kilauea's eruptions, Hawaii continues to grow.

Related Ideas from Across the Atlantic

The next major event leading to the acceptance of the theory of seafloor spreading was a paper published in 1963 by two British geophysicists, Fred Vine and Drummond Mathews. Vine had believed in continental drift since the day that he opened his first geography book. He had been impressed by the map of the South Atlantic that showed the near perfect fit between Africa and South America. In 1962, he was 22 years old and a graduate

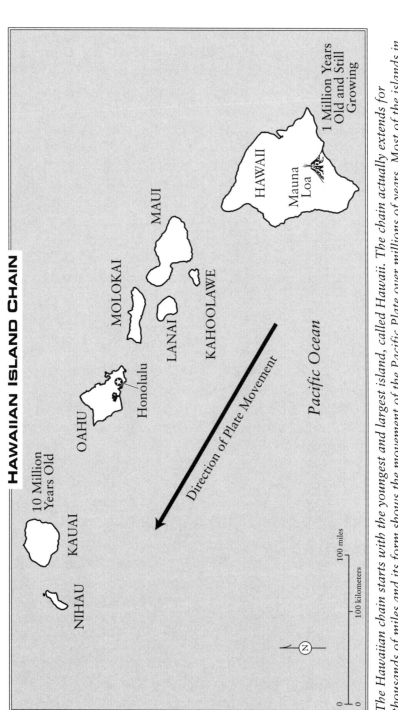

The Hawaiian chain starts with the youngest and largest island, called Hawaii. The chain actually extends for thousands of miles and its form shows the movement of the Pacific Plate over millions of years. Most of the islands in the chain are now submerged guyots, but Midway is one distant member that still shows above the surface.

student of geology at Cambridge University in England. That year Harry Hess visited Cambridge and gave an off-the-cuff talk to an assembly of faculty and students, including Fred Vine. This talk was a preview of Hess's yet unpublished geopoetry. After hearing Hess speak, Vine actively promoted the ideas of the highly respected American geophysicist and focused his own research on furthering those ideas. To this end, he sought out Drummond Mathews, who was also a believer in drift theory, to be his research supervisor. Mathews was only a few years older than Vine. He had just finished his own doctoral program after completing two years of oceanographic survey work in the South Atlantic.

During their research on Hess's theory, Vine and Mathews recognized the importance of the mirror image found in the magnetic stripes on either side of the Juan de Fuca Ridge off the coast of Vancouver Island in the Pacific Ocean.

Although the pattern of stripes was not perfectly clear due to a few seamounts in the area, Vine and Mathews realized that the basic symmetry of the magnetic pattern indicated that the seafloor was spreading on both sides of the ridge. In addition, Vine and Mathews suspected that the zebra-patterned stripes—caused by the alternating areas of greater and lesser magnetic intensity—were actually the frozen history of geomagnetic reversals.

The paper written by Vine and Mathews to clarify the meaning of the unique magnetic stripes was the one published in 1963. Fortunately, the paper had also included some of Mathews's own direct observations on rock magnetism. Without this new information and the backing of his mentor, Vine might not have been able to publish his innovative ideas. Vine's detractors said that Mathews's new information had justified the publication of what was otherwise only empty guesses.

A Less Fortunate Canadian

In Canada, as in Britain, there was a tradition of studying rock magnetism. Lawrence Morley, who received his doctorate from the University of Toronto in 1952, followed the tradition of magnetic studies by writing his dissertation on this topic. His early professional work was on a team conducting a general geological survey of the Canadian Arctic. In 1961, he returned to Toronto as a member of their geology faculty.

Morley then read the papers about the pattern of magnetic stripes found on the ocean floor near the Juan de Fuca Ridge. These articles, published between 1958 and 1961, contained the research findings of the Scripps team of Mason and Raff. By 1962, Morley—totally unaware of the work done by Vine and Mathews—perceived the pattern of magnetic stripes as a sign of seafloor spreading. He quickly prepared an essay covering his ideas and submitted it for publication to the prestigious British journal *Nature*. It was rejected. He made minor revisions and submitted it to another journal. Again, it was rejected. Unfortunately, it did not appear in print until 1964, well after the publication of the paper by Vine and Mathews.

Later, Morley's misfortune was seen in a sympathetic light by most geologists. Indeed, the idea that seafloor spreading is confirmed by the pattern of magnetic stripes is now widely designated as the Vine-Mathews-Morley hypothesis. Morley's lasting fame has been assured.

In the years between 1962 and 1964, however, none of this research was taken seriously in the United States. The hypothesis about seafloor spreading was regarded so lightly that most American geologists joked about the concept at their parties. Soon, however, the geopoetry of another Canadian improved the status of the hypothesis.

Another Scientist Joins the Cause

Like Lawrence Morley, James T. Wilson was a Canadian and a faculty member at the University of Toronto. Early in his career, Wilson discovered that there was already a notable J. T. Wilson in the field of geology. The younger Wilson decided that two J. T. Wilsons were one too many. James T. Wilson was concerned that whatever *he* did would be credited to the other Wilson. He solved the problem by identifying himself by his middle name, Tuzo.

In the mid-1930s, Wilson and Harry Hess were both at Princeton University. Tuzo Wilson had been a graduate student while Hess was an instructor. After he received his degree, Wilson worked briefly for the Geological Survey of Canada. In 1939, he joined the Royal Canadian Army Engineers and later, retired from the Canadian army as a colonel. Wilson was immediately hired as a professor of geology by the University of Toronto. Each summer for the next nine years, Tuzo Wilson was in the field, studying geological formations in central Canada and the Canadian Arctic. He was also involved with the workings of scientific organizations and government agencies on both sides of the border. In the early 1960s, Tuzo Wilson stopped studying geological formations in the field and began to theorize about them.

Tuzo Wilson was not the first geologist to speculate that the great crustal plates were the main features of the earth's surface. He sensed that the observations of geophysicists like William Menard were related to the idea of crustal plates and the possible movement of such plates. At first, Wilson had been puzzled by the long cracks that Menard had been the first to identify. He assumed that weak areas in the crust caused the cracks to form *before* ridge-rift structures developed. He came to believe that as the rift formed, its course would, at times, be offset by these "protocracks." Consequently, rifts might not follow a perfectly straight line. This idea turned out to be correct. Rift lines are often displaced by lateral faults.

Wilson's ideas resulted from his study of a specific location. He observed that the geological formation that includes the famous San Andreas Fault in California consists of three features. This formation begins off the coast of Panama as a long, deep ocean trench. As the result of some enormous, long-ago upheaval, the northern portion of the ocean trench was pushed beneath the continent along the west coast of California. After the burial was completed, the trench was transformed into an underground fault. The deeply buried fault continues below the surface of California for many miles. Its path is now marked by long cracks in the surface of the continent. Finally, the fault line emerges from under the continent just north of San Francisco. It proceeds for another 500 miles (800 km) and is transformed into the Juan de Fuca Ridge. This ridge-rift is detectable on the floor of the ocean near Vancouver, Canada. Wilson believed that the San Andreas Fault linked the deep ocean trench off the Central American coast and the ridge-rift off Vancouver in Canada. In brief, according to Wilson, elements of this geological formation had been transformed from an ocean trench to a fault and finally to a ridge-rift.

To further his hypothesis, Tuzo Wilson chose to designate Menard's "fracture zones" by a new name, "*transform* faults." As might be expected, William Menard was not happy with this name change. From then on, he felt that his contribution to earth studies had been slighted.

The ideas of the geopoets can be quickly summarized. Hess and Dietz independently theorized that the great crustal plates are moved by the process of convection caused by the rising magma along the ridge-rift structures. Vine, Mathews, and Morley saw the pattern of magnetic stripes on the ocean floor as the record of magnetic field reversals. They also proposed that the mirror symmetry of these stripes was an indication that the ocean crust was being carried away from both sides of the center rift. Wilson believed that he discerned the relationship between ridge-rifts, faults, and deep ocean trenches. He recognized them as boundaries of the vast plates that made up the surface of the earth.

9
A Bit of Geometric Spice

Most scientists are fond of numbers. If an idea can be expressed in mathematical terms, they believe that the idea becomes "elegant." Mathematics lends a sense of precision to a set of ideas, and the ideas become more acceptable.

Sir Edward Bullard, a senior British scholar at Cambridge University, was the first to provide the inclusion of mathematics to the brewing controversy about the theory of continental drift. Bullard, like Hess and Ewing, had been influenced by Professor Field. All three men looked to physical oceanography for clues about the earth's history.

Bullard came from a prosperous family and did not need to worry about his future. He studied physics at Cambridge University in the late 1920s and early 1930s, but his scholarship was far from brilliant. Although his undergraduate thesis (extensive research paper) was accepted, Bullard was told by Lord Rutherford, then Britain's most prominent physicist, that he had no future in the field. Consequently, he changed his specialization to geology.

Bullard began his career as a geological field technician in Africa. He served in the Royal Navy during World War II, taught for a short time at the University of Toronto, and spent the summer of 1949 at Scripps with William Menard. Bullard then worked for five years as a government official in London before returning to Cambridge as a professor of geophysics.

In 1964, two years before the breakthrough at Lamont, Bullard spoke to geologists at a special meeting of the British Royal Society in London. He reintroduced the concept of the jigsaw puzzle fit between the boundaries of Africa and South America. In effect, he was restating Wegener's ill-fated ideas. Bullard's presentation, however, had three advantages. He utilized a 200-year-old spherical geometry theorem as a research tool, employed a state-of-the-art computer, and used his flair for dramatic delivery.

The theorem had been devised in 1760 by a Swiss mathematician named Euler. It states that when a segment of the surface layer of a sphere is moving, the segment tends to rotate around one fixed point. In other words, one point of the moving segment tends to remain immobile while the rest of the segment swivels around that point. Therefore, it would be incorrect to assume that Africa and South America had drifted away from one another in a strictly westerly or easterly direction. Rather, they rotated as they moved.

Indeed, the result of Euler's Theorem can be demonstrated by looking at a map of the world. When the continent of South America *is* rotated a few degrees counterclockwise and that of Africa is rotated clockwise, the fit between the two is nearly perfect.

By 1964, the year that Bullard made his presentation, most geologists realized that the fit between Africa and South America should be made using the ancient continental boundaries rather than the present coastlines. Therefore, when Bullard displayed his mathematical computations—calculated on his modern computer—and dramatically unveiled his prepared map, the scientists were intrigued. Indeed, the few areas of mismatch were seen as insignificant compared to the thousands of miles of neatly joining continental boundaries. The belief that this close fit was accidental or coincidental could no longer be accepted.

On that memorable occasion in 1964, Bullard's audience was made up of members of the British Royal Society. In 1926, when

Wegener's ideas were debated in New York, the scientific community was not receptive to innovative theories of geology. Forty years later, Bullard's refinement of the continental drift theory was received far more readily. The scientific community in Britain began to accept the theory that Africa had once been joined to South America and that the two continents had slowly separated.

Bullard's presentation introduced the benefits of a mathematical approach to the whole field of geology. The next step was to determine how to use these mathematical tools to verify other scientific observations. Some geophysicists hoped that Euler's geometry would generate a better comprehension of the position of the continents on the face of the earth. Until the geological implications of Euler's geometry were shown to be more widely applicable, the theory of continental drift was still incomplete.

After Bullard's work in 1964, other developments rapidly followed. Crucial observations were made in 1965 and 1966. In late 1966, the theory of seafloor spreading was accepted. However, another year would pass before the complete picture was revealed. In 1967, two groups of young scientists, working independently, developed conclusive mathematical arguments. Their insights allowed geologists to integrate the complete array of observations into the theory of plate tectonics.

10

Upstarts at Lamont

While some scientists worked to substantiate geological theories with the use of mathematical theorems, others continued to conduct research and develop ideas about the theories of continental drift, seafloor spreading, and plate tectonics. Each concept was discussed, debated, and ofttimes ridiculed by the scientific community. Most young geologists agreed that the concepts about seafloor spreading were pertinent to their proposed revision of Earth's history.

As late as the mid-1960s, seafloor spreading was still not widely accepted by the majority of geophysicists and even less well accepted by traditional geologists. Almost all agreed that sufficient evidence was still lacking. This evidence finally emerged in the period between late 1965 and late 1966.

During those months, key components of the theory of seafloor spreading were confirmed and documented. These factors were uncovered by a group of graduate students—sometimes called "upstarts" by traditional geologists. The young people worked in basement laboratories and offices of the Lamont Geological Observatory and on the main campus of Columbia University.

Their research resulted in three main discoveries. First, a pattern of magnetic reversals was found in cores of sediment taken from the floor of the Atlantic Ocean. These cores had been stored for many years in the basement of Lamont. A short time later, computer analysis of new underwater recordings revealed identical magnetic patterns on each side of the Antarctic Ridge-Rift structure. Lastly, a new analysis of existing information about undersea earthquake events confirmed Tuzo Wilson's concepts of subductive zones and the recycling of the oceanic crust. The stories behind each of these breakthroughs demonstrate how hard work, good scholarship, and sheer luck play a part in every scientific revelation.

In the mid-1960s, Maurice Ewing, then director of Lamont Geological Observatory, convinced his former student Neil Opdyke to return to Lamont as a project director. In 1959, Opdyke, while a Lamont undergraduate, had made a name for himself when he assisted Runcorn in resolving the problem of the "wandering" magnetic poles.

Ewing hoped that Opdyke could assist with another problem. Over the years, Lamont physical oceanographers had brought back numerous samples of undersea sediment. The long, carefully labeled samples, or cores—which measured from 10 to 30 feet (3–9 m)—were taken from various parts of the Atlantic seafloor and kept in basement storerooms. Aside from dating some of the cores by their fossil remains, no one had bothered to assess the scientific worth of the collection. Ewing now wanted Opdyke to analyze these cores for remanent magnetism and do other analyses of their properties.

Opdyke's view of this project differed sharply from that of Ewing. The younger man thought that studying the cores was probably a waste of time and money. However, he accepted the project and then requested John Foster, a graduate student, to conduct the necessary testing. Foster was not too enthusiastic, either. He was aware that no one at Lamont had ever obtained a magnetic reading from a core of seafloor sediment. However, the young scientist hoped that a very fine-tuned magnetometer might possibly achieve the desired results. Foster set out to design a new and better instrument that would help analyze other rock specimens, even if it were unable to obtain readings from the cores.

In late 1965, shortly after Foster had completed his new magnetometer, Bill Glass, a fellow graduate student, asked for a demonstration of the untried instrument. Glass had brought in a sample from one of the cores to use for the demonstration. Neither Foster nor Glass expected any response from the machine but, amazingly, the instrument revealed a magnetic response. Foster and Glass tested more samples. Their luck held! They were consistently obtaining clear patterns of magnetism. In order to date their results, the geologists tested samples from a core that had been previously analyzed for fossil remains. Comparing the results from fossil records and the magnetometer readings, they were able to assign dates to the magnetic patterns on the samples.

In addition, they found something truly amazing. Most of the samples registered the normal pattern of magnetic orientation toward the North Pole, but one showed *reversed* magnetic orientation—toward the South Pole. They had accidently chosen a sample that supplied new information about reverse polarity.

Six years before, Ronald Mason and Arthur Raff from Scripps Institution of Oceanography, had recognized the same startling patterns of normal and reverse polarity. They had found the patterns in red clay cores from the floor of the Pacific off the coast of Mexico. When the two men reported the discovery to their supervisors, they were told that their findings were incorrect, silly, or too fanciful for publication. The two geophysicists did not tell anyone else. Now, however, the world of science was better prepared to accept new information on reverse polarity from cores of sediment.

After Foster and Glass reported their findings to Opdyke, the older man immediately recognized the potential significance of their discoveries. The students also phoned Glass's research adviser, Heezen. He, too, was intrigued by the news.

Without delay, Foster and Glass began the painstaking procedure of sample preparation and sample testing. They were now working under Opdyke's direct supervision—with Heezen looking on. The two young men conducted test after test. They analyzed each core's pattern of magnetism from the most recent time–found in the uppermost part of the sample–to the most ancient–located at the bottom. Testing proved that the patterns of polarity reversals were the same from one core to another regardless of their original location in the seafloor. Opdyke, Foster, and Glass had discovered a way to measure time in geological epochs. They cross-checked their observations with those that the Geological Survey scientists had found in volcanic lava samples from dry land. The general pattern looked very similar. Slowly, pieces of the puzzle were beginning to fall into place.

In January of 1966, another Lamont graduate student, Walter Pitman, was carefully analyzing records in his basement office across the hall from Opdyke's. His information had been collected by the research ship *Eltanin* from a magnetometer towed a short distance above the ocean floor. A few weeks earlier—in the Southern Hemisphere's summer—*Eltanin* had made several crossings of the ridge-rift around Antarctica. Pitman, with the help of another graduate student, Ellen Herron, had directed the magnetic study.

After Pitman and Herron fed their information into a Lamont computer, they received printouts with graphic profiles of the seafloor's magnetic intensity. The two students carefully analyzed records from two different crossings of the ridge-rift—coded *Eltanin*-20 and *Eltanin*-21. For purposes of comparison, they then processed information from a previous crossing—identified as *Eltanin*-19.

Pitman and Herron looked at the newer profiles and then turned their attention to *Eltanin*-19. They saw something truly electrifying. The graphic representation of magnetic intensity on one side of the ridge-rift was an almost perfect mirror image of the magnetic pattern on the other side.

The profiles reminded Pitman of the research done by Vine, Mathews, and Morely on the strange zebra-patterned stripes located at the Juan de Fuca Ridge. Unlike the earlier observations, however, those collected by Pitman and Herron were not

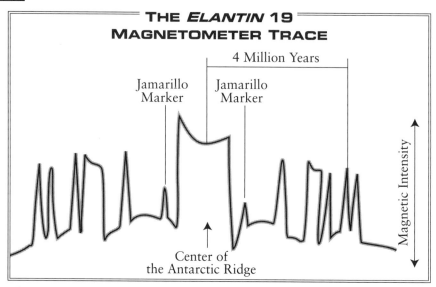

This is a record of magnetic intensity across the Antarctic Rift. It was so symmetrical that skeptics thought it could not be real.

distorted by other seafloor features such as seamounts. The nearly symmetrical patterns that registered on each side of the rift were crystal clear.

Although the widths of the stripes differed, the zebra pattern recorded at the Antarctic Ridge closely matched the pattern found at the Juan de Fuca Ridge. The students saw the variation in widths as an indication that the rate of seafloor spreading differed from rift to rift and from time to time. They reasoned that if similar magnetic patterns were recorded at ridge-rifts located thousands of miles apart, these patterns might be found at every ridge-rift around the globe. This would confirm the theory that the seafloor was spreading at every ridge-rift.

The same day that Pitman realized the significance of the matching profiles, he and Herron spent hours checking and rechecking their information. Late that night, Pitman pinned the profiles of *Eltanin*-19, -20 and -21 to Opdyke's door. This event helped initiate the geological revolution of 1966.

When Pitman returned to Lamont the next day, Opdyke was quite excited. He knew immediately what the profiles meant.

Other geologists doubted the truth of the discovery. Some complained that the *Eltanin*-19 profile was "too perfect" to be believable. Later, the disbelievers were proved to be wrong.

However, there *was* cause for some doubt and confusion. During the complicated validation procedures, different magnetic profiles were sometimes incorrectly aligned. A short period of normal polarity, the Olduvai Event, was confused with another short period of normality—previously unnoticed by Pitman and Herron. Soon, they realized that they were looking at not one but two, brief, normal periods. In computer-generated profiles, these short stretches of normal polarity were recognizable changes within the longer reversal epoch. The Lamont scientists were thrilled to have discovered what they thought was an additional short normal interval. They even gave it a name.

In reality, this interval—now called the Jaramillo Event—had been observed a few months earlier by Cox and his colleagues at Menlo Park. The people at Lamont, however, had not been aware of the finding. This misconception was not cleared up until a meeting in April of 1966.

Soon, the news about the *Eltanin*-19 profile was circulating around the offices at Lamont. Lynn Sykes from the Geology Department at Columbia University's main campus had also heard the exciting announcement. Indeed, he and his coworkers were inspired to discover something equally important. They busied themselves analyzing earthquake records from an area east of Japan. The earthquake centers in this region were known to lie 50 miles (80 km) below the floor of the Pacific. The analysis of sound waves indicated that the earthquakes were being caused by the movement of a 5-mile (8 km) thick slab of rock. The slab seemed to be tilted at a 45° angle and buried under the front edge of the deep trench that runs southward off the east coast of Japan. It looked as if Lynn Sykes and his fellow students had located a place where old ocean crust begins its descent into the magma layer. Tuzo Wilson's idea about recycling old crust at the subductive zones was being confirmed.

In April of 1966, a meeting of the American Geophysical Union was held in Washington, D.C. All the important research people were present. Included among the members were Opdyke, Foster, Pitman, and Sykes from Lamont and the Cox team from Menlo Park. Of course, some of the main topics of discussion were the recent discoveries at Lamont. Much information on these topics was exchanged, but it was obvious that gaps still existed in their arguments. More research and deliberation was needed.

However, this open exchange cleared up at least one important misunderstanding. The Lamont team had proudly announced that they had detected a previously unknown period of normal polarity within a long era of reverse polarity. The geologists at Menlo immediately made it very clear, however, that they had already uncovered that event—several months before. Indeed, they knew it as the Jaramillo Event. Although disappointed, the Lamont scientists were excited by the newly established agreement about the significance of the find.

After that April meeting, all the scientists rushed back to their laboratories to complete their research and refine their interpretations. Fred Vine, the young British geologist, was especially eager to return to his research at Princeton University. In 1963, he and Drummond Mathews had achieved some notoriety with their article on seafloor spreading. Although Vine's ideas had met with little approval, he was still deeply involved in studying that topic. In the months before the meeting, he had spent large amounts of time analyzing magnetic reversal patterns from every available seafloor survey. After the meeting, Vine followed an even more demanding schedule of research because he wanted new information to refute his many critics. Vine firmly believed that the theory of seafloor spreading eventually would gain universal approval. He wanted to make sure that he, Mathews, and Morley were given credit for their part in that acceptance.

The next act in the drama came a few months later, in November of 1966. The National Aeronautics and Space Administration (NASA) sponsored a meeting that was held at Columbia University. Since the important meeting in June, geologists at the main campus in New York City and

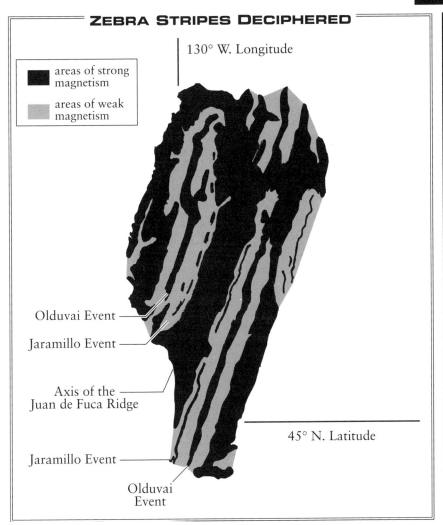

Once scientists knew what to look for, they could see the same pattern of symmetry at the Juan de Fuca Ridge that had been found at the Antarctic Ridge. Once seen, the pattern confirmed what Vine, Mathews, and Morley had said years before.

at the Lamont facility on the Hudson River had been working night and day.

Attendance at that November meeting was very selective. Fewer than 40 people had been invited. Included in the group were a number of highly influential scientists and, of course, several of the younger geologists. The established scholars were assembled to hear all the arguments about the theory of seafloor spreading. They would decide the fate of Wegener's concepts—proposed 40 years before.

As the story goes, the proceedings of that meeting finally convinced William Menard to accept the theory. His conversion can be regarded as a milestone because he was the perfect person to judge the validity of seafloor spreading. For years, Menard had been a neutral observer in all the discussions. He was familiar with the pertinent facts, had reviewed the theory papers, and knew how the observations had been collected. Menard's acceptance signified that the revolution in geology had been won. The finishing touches might take some years, but the outcome was never again in doubt.

A remarkable feature of the revolution was the youth of the main participants. Opdyke, although a supervisor, was only a few years older than the students, Foster, Pitman, and Herron. Sykes was not quite 30 years of age in 1966. These young people were part of a second generation—the direct descendants of pioneers such as Ewing and Hess. The older scientists had gained experience during World War II, and then designed new sets of research instruments, procedures, and goals. Indeed, they had instructed a new generation in how research should be conducted. The younger people followed their advice—and with hard work, sound research, and good fortune they transformed the study of geology.

II

The Adventure Continues

During the years between 1946 and 1966, only a handful of scientists were thinking about continental drift. The large majority of geologists focused their attention on other projects such as locating deposits of oil and other valuable minerals on the continents or under the seas. Many research geologists spent their time preparing maps that indicated the distribution of volcanoes, earthquakes, or rocks of different types and ages. Some studied the effects of glaciers on the contour of the land. Others investigated techniques and equipment for finding and refining oil. Any link between these projects and the new theories in geology seemed slight.

Of those American scientists who thought about continental drift, most were neutral observers of the ongoing arguments. The passionate attacks on Wegener's ideas now appeared extreme. A few of the traditional geologists must have been aware that the growing body of evidence indicated some form of continental movement. However, even those scientists felt no urgency to evaluate the new ideas.

Outside the United States, there was still some tendency for argument. In England, many geologists were favorably disposed toward the idea of seafloor spreading. They had studied the work of Arthur Holmes. His textbook *Principles of Physical Geography*—revised in 1945—supported the general idea of continental drift. Most English geologists respected Holmes and believed that the idea deserved consideration if not acceptance. However, Sir Harold Jeffreys, another well-regarded English geologist, was an opponent of these ideas. Jeffreys taught that continental movement was impossible because the motion of the vast plates would create an enormous amount of friction. Other strong opponents to the ideas were at work in Russia, Germany, and elsewhere.

In the United States, some geologists were ready to consider new ideas on continental drift if these ideas were presented with new evidence. Such evidence had been revealed at the NASA meeting of November 1966. However, only 40 scientists had been present at the meeting. The remaining 30,000 or more earth scientists remained uninformed or little informed about new research on continental drift. This new information spread slowly through the scientific community, and worldwide acceptance took several years. In fact, 10 years passed before the complete theory of plate tectonics was presented in introductory textbooks.

A crucial step in achieving wide-spread acceptance was the inclusion of a strong mathematical foundation. William Bullard had begun the process by applying the use of Euler's geometric theorem to explain the present locations of the earth's land masses.

Two groups of young research scientists—influenced by advocates of the new geological theories—set out to expand the application of Bullard's work and Euler's theorem. These people realized that the use of spherical geometry would add elegance, and possibly total acceptance, to the evidence that was being collected.

One of the young American scientists was W. Jason Morgan. He had received his Ph.D. in physics from Princeton University in 1964. In early 1967, Morgan read an article by William Menard in the journal *Science*. Menard had compared fracture zones found in the Pacific Ocean with those found in the Atlantic. His illustrations showed that all the fracture lines were shallow curves, not straight lines. Jason Morgan observed that the degree of curvature slowly increased as the distance from the rift increased. Morgan realized that the path of these gently curved lines might be duplicated in a mathematical model using Euler's theorem. By adding new information about the lines of fracture at the Mid-Atlantic Ridge, he showed that the Euler geometric calculations would allow him to predict the path of the fractures illustrated in Menard's article.

A few weeks later, in April of 1967, Morgan was scheduled to give a talk at a meeting of the American Geophysical Union. At the last minute, he decided to change the topic of his lecture. Morgan was so enthusiastic about his use of Euler's theorem that he wanted to tell other scientists about his research as soon as possible. Unfortunately, it was too late to notify anyone of his decision. Morgan's audience, expecting another subject, was ill-prepared to understand the implications of his intricate mathematical model. The lecture received few comments.

Dan McKenzie, another geologist at the meeting, would have immediately understood Morgan's work. However, McKenzie, a graduate student from Cambridge University in England, attended a different lecture.

For many months before the meeting, McKenzie, too, had been preparing a mathematical model—also using Euler's Theorem—to describe the path of fracture lines. In late 1966, he had finished his complicated calculations with the help of Robert Parker, a young mathematician working at Scripps. McKenzie submitted his article to *Nature* at the end of June 1967, unaware that Morgan was also working with Euler's Theorem. McKenzie's paper was published in December 1967. A few months later, in March 1968, the print version of Morgan's talk appeared in the *Journal of Geological Research*.

The paper published by McKenzie and Parker was the first to define the theory of plate tectonics in a fully scientific and mathematical manner. However, Morgan's lecture on the same subject had taken place nine months earlier. After his paper was published, it was regarded by many as the most important geology paper ever written. Certainly, Morgan's work—along with that of McKenzie and Parker—helped to expand the number of believers. More and more people became convinced that the theory of plate tectonics explained the known facts and pointed the way for new investigations.

By now, the terminology of the new theory had changed to reflect the differences between Wegener's original ideas and the more recent developments—such as the use of spherical geometry. Geologists now spoke about the plate tectonic theory instead of the theory of continental drift or seafloor spreading. Wegener had seen the continents as great ocean liners plowing through the oceanic crust. Scientists had found this concept very difficult to accept. Among other problems, they knew that the energy required to move these huge land masses would have been enormous. Scientists also reasoned that many fragments of oceanic crust would have been left behind as the continents plowed through the crust. Neither the probable source of that tremendous energy nor the absence of fragments of crust were ever treated in a convincing manner by Wegener.

More Converts

A workable theory generates attention and acceptance by many scientists—but not all. Frequently, nonacceptance results from a lack of communication between different branches of the same basic field of science. Geophysicists and physical oceanographers may read different scientific publications, attend different meetings, and have very different research interests. Not all are equally aware of the various developments in neighboring fields. As an example, by 1967, most physical oceanographers knew that their research interests were being helped by the introduction of mathematics. However, many traditional geologists were unaware of the innovative use of mathematics in geological research.

In order for a new theory to gain acceptance by the majority of scientists, each scientist must not only understand the theory and accept its validity, but also they must find it useful in guiding their own research.

Scientists relate many stories about the discovery of useful applications of the plate tectonics theory. One episode began about 1959 when massive seams of coal were discovered in Antarctica. This find was confusing because coal is the remains of plant life. However, plants such as grass, trees, and mosses do not now grow in Antarctica.

A few years after the discovery of coal in Antarctica, a geologist from New Zealand, Peter J. Barrett, found a piece of bone in some stream gravel about 75 miles (120 km) from the coal seams. Barrett thought the bone was probably the lower jaw of an extinct amphibian. He sent the fragment to Dr. Edwin H. Colbert, curator of fossil reptiles and amphibians at the American Museum of Natural History in New York City. Colbert confirmed Barrett's tentative identification and noted that this was the first ancient fossil bone found in Antarctica.

In the winter of 1969, summer in the Southern Hemisphere, Colbert conducted an expedition to pursue Barrett's find. The research group camped near the coal beds. Before they could set out for the site of Barrett's discovery, the weather turned bad and kept their explorations near the camp. This was a stroke of good luck.

On December 4, James Jensen, one of the team's geologists, discovered a piece of jaw and a tooth near their camp. These were identified as belonging to the family of *Lystrosaurus*, a group of mammallike reptiles that had been found in South America, Africa, and Australia. Soon, many more finds of similar importance were discovered by other members of the expedition. Geologists reasoned that the animals must have found their way to the now frigid Antarctica when it was part of an ancient supercontinent and much nearer to the equator. These discoveries confirmed Wegener's long-disputed proposal about the existence of supercontinents. Traditional geologists now perceived that the theory of plate tectonics explained such strange finds and some of the basic problems in geology.

Many years after the revolution of the mid-1960s, the theory of plate tectonics has opened up many new avenues of investigation. Geologists are seeking new information about the ancient history of the earth and the formation of mountains and other large features. The most intriguing new direction is the study of the depths below the continental masses. Very deep drilling (over 10 miles [16 km] down) and powerful new methods of echo sounding, recording, and interpretation are making another part of this planet perceptible to scientists. No one may ever see the deep bowels of the earth, but soon, people may know much more about its mysteries.

The coming generations of earth scientists will be confronted by other mysteries. In meeting these challenges, they will be privileged to follow in the footsteps of Wegener and all the other adventurous scientists who made a revolution in understanding the forces of nature.

Glossary

abyssal plain: Relatively flat area of the ocean floor at a depth of one or more miles (1.6 or more km).

anomalies: Some very unusual occurrences that are not easily explained.

asthenosphere: The second layer of the earth's mantle. It is believed to be rock that is melted to form a thick liquid.

atom: The building block of chemistry; the smallest particle that can be identified as a specific element.

basalt: The principal rock material in molten lavas and magmas.

climatology: The study of large-scale and long-lasting conditions of the atmosphere.

conductivity: The property of a material that permits the material to transmit energy such as heat or electricity.

continent: One of the six major landmasses (Africa, Antarctica, Australia, Eurasia, North America, and South America).

continental crust: The thick upper layer of the earth that rises above the level of the seas and that is composed mainly of granite.

continental drift: The idea that the continents have moved relative to one another over geological time.

continental margin: The broad area at the edge of the continents that includes the seashore, the continental shelf, and the continental slope.

continental shelf: The edges of the continental land mass that are below sea level. The shallow shelf generally extends about 50 miles (80 km) from the visible seacoast.

convection: The movement of any fluid medium in a circular pattern where the movement is driven by a heat source and the movement direction is from the hottest location toward a cooler location.

core: The central part of the earth—about 2,000 miles (3,200 km) in diameter. The outer core appears to be liquid, and the inner core appears to be a solid ball.

dome: A spherical roof or lid; the rounded shape of a small area the earth's crust that is forced upward by rising magma.

dynamo effect: The generation of a magnetic field by an electric current.

earthquake: Short, rapid movements of portions of crust that release the stresses that are brought about by the continuous but slow movements of the great crustal plates.

echo sounding: The use of artificial sound to determine distances by timing the echo.

Euler's Theorem: The idea that the movement of any segment of the surface of a sphere tends to follow a curved path.

fault: A crack that separates blocks of the earth's crust.

fossil: The naturally preserved evidence of once-living creatures including teeth, bones, and shells.

fracture zone: Large, long, slightly curved cracks in the ocean floor—often found at right angles to ridge-rift features.

geographic pole: The two points—North and South—around which the earth rotates.

geography: The study of the distribution of physical features, animals, vegetation, minerals, and human habitation on the earth.

geomagnetism: The magnetic field that covers the earth.

geophysics: The study of the physical properties of the earth.

granite: The principal rock of the continental crust. It is rock that was once melted, and it contains coarse grains of glasslike particles. It can be polished to a high lustre and its often used in the facades and pillars of public buildings.

guyot: A seamount of generally conical form with an upper surface flattened by surface wave erosion.

igneous rocks: Rock that was once molten and has since cooled and solidified.

Jaramillo Event: The relatively short period of normal magnetism near the end of the most recent period of polar reversal.

jet stream: A broad and fast-moving stream of air that flows at an altitude of more than 5 miles (8 km).

lava: Molten rock that has erupted onto the surface of the earth.

limestone: Sedimentary rock that is composed mainly of calcium compounds that once formed the shells of small sea creatures.

lodestone: Rock having a relatively strong natural magnetic field.

magma: Molten rock beneath the earth's surface. If and when it penetrates the earth's surface, it becomes lava.

magnetic field: The pattern of energy that surrounds a magnet.

magnetic orientation: The direction in which the poles of a freely moving magnet come to rest with respect to the earth.

magnetic poles: The two points on a magnet where the magnetic force is strongest.

magnetic reversal: A dramatic change in the earth's magnetic field. The south magnetic pole moves to the location of the north magnetic pole and vice versa.

magnetic stripes: Alternating sections of the ocean floor that have opposite magnetic orientations.

magnetism: The property of an object for attracting or repelling other objects.

magnetometer: A machine that detects a magnetic field and reveals its orientation.

mantle: The deep layer of rock between the earth's crust and its core. Most of the earth's material is incorporated in the mantle.

metamorphic rock: Rock that has been changed from its original form by heat, pressure, or chemical action.

meteorology: The study of weather.

oceanic crust: The relatively thin (3 to 6 miles, 4.8 to 9.6 km) layer of basaltic rock that makes up the floors of the oceans.

oceanography: The study of the physical, chemical, and biological nature of the oceans and the ocean floors.

Pacific rim: The boundary of the Pacific Ocean—often used to designate the countries that face onto the Western Pacific.

passive margin: A boundary between the ocean crust and the continental crust where both are part of the same plate.

physical oceanography: The study of the physical properties of the ocean and the ocean floor.

plate, crustal: Very large sections of crust mainly composed of basalt and which may include the continents that are mainly granite.

plate tectonics, theory of: The idea that all the large segments of the earth's crust are in constant motion.

plume: A featherlike object; specifically, the form of large streams of magma rising within the asthenosphere to the region just beneath the earth's crust.

polarity epoch: The long span of time during which the magnetic poles remain in one location.

polar wandering: Apparent movement of the magnetic poles relative to the continents.

productive zone: The slowly separating boundary between two crustal plates where new crust is being formed.

radioactivity: The property of giving off invisible but highly energetic particles.

radiometric dating: Estimating the age of a rock sample by measuring the proportions of materials that have changed by releasing radioactive particles over time.

remanent magnetism: The magnetism taken into rock from the earth's magnetic field.

reverse polarity: A condition such that the earth's magnetic poles are in the locations opposite to the locations now occupied.

ridge-rift feature: Long seams in the ocean's crust that mark the boundaries between plates. The ridges that border the seam are separated by the rift valley that can be as much as 100 miles (160 km) across.

rift: A large and lengthy valley in the ocean floor—often called a gorge when present on a continental land mass.

seafloor spreading: The idea that crustal plates gradually separate at the site of the oceanic rifts and that new crust is formed in the gap by rising magma.

seamount: A mountain that arises from the floor of the sea.

sedimentary rock: Rocks formed from compressed deposits of silt, clay, sand, or other fine-grained materials including the small shells of sea animals.

seismic: Vibrations in the earth caused by earthquakes or other concussions such as explosions.

seismograph: An instrument for detecting and recording vibrations in the earth.

solar system: The sun and its planets and the other, smaller bodies that move in orbits around it.

sonar: The system that generates sound waves underwater and that detects and records the echoes that are formed when the sound waves strike an underwater object.

strata: Layers of rock.

stratification: The buildup of a series of layers—one on top of the other.

subduction: The movement of an oceanic plate in a downward direction under a continental mass. The descending front of the plate typically reaches a depth of 50 miles (80 km) or more before extensive remelting takes place.

supercontinent: The assembly of all the continents into a single mass.

tectonic: Movements of portions of crust that change the contours of the land or ocean floor.

topography: The shape and form of the earth's surface.

transform fault: A fracture in the crust along which lateral movement of crustal segments can take place. The San Andreas Fault in California is an example.

trench: A deep trough in the ocean floor thought to be the result of the descent of ocean crust beneath the edge of a continent.

triple junction: A geographic location where three crustal plates come together.

turbidity current: A flow of mud, silt, sand, and gravel along the floor of an ocean that is generated by an earthquake centered within the continental margin.

uranium: A heavy metallic element that is naturally radioactive.

volcanic hotspot: An area where hot magma reaches upward so that it is just beneath or penetrating through the crust.

volcanism: The processes by which material is transported from the earth's interior to its surface.

Further Reading

Erickson, Jon. *Plate Tectonics*, New York: Facts On File, Inc., 1992. A short, readable, and straightforward treatment of the concept of plate tectonics and the scientific implications of the theory.

Glen, William. *The Road to Jaramillo*, Stanford, Calif.: The Stanford University Press, 1982. Provides a detailed historical review of the work of Allen Cox and colleagues in the development of the science of geomagnetism and the study of magnetic reversals.

Menard, H. W. *The Ocean of Truth*, Princeton, N.J.: The Princeton University Press, 1986. A narrative history of the activities conducted by the staff of the Scripps Institution of Oceanography with additional coverage of the work done by the research people at the Lamont-Doherty Geological Observatory that led to the final acceptance of the theory of plate tectonics.

Schwartzbach, Martin. *Alfred Wegener*, Madison, Wisc.: Science Tech, Inc., 1986. The definitive biography of the "father" of continental drift theory.

Vogel, Shawana. *Naked Earth*, New York: Dutton Books, 1992. A brief, simplified account of the factual basis for the theory of plate tectonics.

Wertenbacher, William. *The Floor of the Sea*, Boston, Mass.: Little Brown and Co., 1974. A complete historical review of the career of Maurice Ewing and the activities at the Lamont-Doherty Geological Observatory.

Index

Italic numbers indicate illustrations.

A

abyssal plain 53–55, 81
Aden, Gulf of 94
African Plate 5
Alaska 14
American Association of Petroleum
  Geologists 38
American Geophysical Union 111–112,
  117
American Museum of Natural History
  119
Anatolian Plate 2
Andes Mountains 10
Antarctica 119
Antarctic Plate 16
Antarctic Ridge-Rift 107, *110*, 113
Appalachian Mountains 12
argon gas 75
Aristotle 25
Arizona State University 96
Assal Rift, Africa 94
asthenosphere 4–6, 8, 11, 92–94
Atlantis (research vessel) 51–53
atom bomb 80
Atwater, Tanya 89
Australian National University at Canberra
  74
avalanches, underwater 54–55

B

Bacon, Francis 31
Baird (research vessel) 83
ballooning 33
Barrett, Peter J. 119
basalt 2, 10–12, 93
Bible 25–28
Bikini Atoll 80
biology 17
Bowie, William 50
British Royal Society 104

Brown, Mike 56
Bullard, Sir Edward 103–105, 116

C

Cambridge University 48, 70, 99,
  103–104, 117
Cape Cod, Massachusetts 45
Chamberlin, Rollin 38
Clark Air Force Base ix
Climates of the Geological Past, The
  (Koppen and Wegener) 37
coal 119
Cocos Plate 2
Colbert, Edwin H. 119
cold war xii
Columbia University 48, 51–52, 61, 80,
  106, 111–112
comets 32
compass 67
computers 16, 104
continental boundary 10
continental drift v, x–xi, 3, 12, 21,
  43–45, 49, 61, 66, 73, 76, 82
  Bullard's support of theory of
    103–106
  Holmes's thoughts on 115–116
  revision of theory of 118
  Wegener's idea on 32, 34,
    37–39
continental shelf 45, 50
convection 6–7, 8, *13*
core, Earth's 4
cores *See* plugs
cosmic rays 61
Cox, Allan 63–67, 73–76, 78–79,
  111–112
crust, Earth's 2–3, 4–8, *15*, 29, 52, 59,
  81, 101, 107, 111
  formation of 10–*13*, 14
  movement of 93–96

D

Dalrymple, Brent 63, 67, 74–76, 78
dating rocks 23, 27–28, 63, 69–70, 74–75, 76, 78
diamonds 27
Dietz, Robert 81–82, 90–93, 96
Doell, Richard 63–67, 73–76, 78
dome, magma 7, 12, 14

E

earthquakes x, 1–3, 14, 19, 26, 29, 54, 58–59, 107, 115
echo sounding 17, 30, 36, 45–46, 57, *59*, 64
    method of *19–20*
    oceanic 50–53
    use of in Pacific Ocean 81–82
Eckart, Carl 80
Elastic Waves in Layered Media (Ewing, Jardetzky, and Press) 62
Eltanin (research vessel) 109
Eltanin-19 109–111
Eltanin-20 109–110
Eltanin-21 109–110
Euler, Leonhard 104–105
Euler's theorem 104–105, 116–118
Eurasian Plate 2
Everndon, Jack 74
Ewing, John 56
Ewing, Maurice 59–63, 80–81, 83, 91, 103, 107, 114
    early work 44–45, 49–53
    expeditions of 56–57

F

faults *8*, 14–*15*, 29, 82–83, 101–102
Field, Richard M. 44–45, 50–51, 91, 103
fieldwork 17, 49
fossils 23, 25–26, 28, 32, 37, 107–108, 119
Foster, John 107–109, 112, 114
fracture zones 82–84, 89, 93, 116
Fulbright Fellowship 91

G

gems 30
Geological Survey of Canada 101
geomagnetism 72 *See also* magnetism
geophysics 17, 19, 21, 64–65
geopoets 90, 99–100, 102
Georgi, Johannes 40
German Marine Observatory 37
Glass, Bill 107, 109
global expansion 59–60
Grand Banks 54
granite 2, 12, 93
gravity 21, 52
Great Rift Valley, Africa *8*
Greenland 14, 16, 33–*35*, 36, 39–41

Guyot, Arnold Henry 92
guyots 92–*94*, 95, 97–98

H

Hawaiian Islands 3, 48, 97–*98*
Heezen, Bruce 56–62, 83, 108
Henry L. and Grace Doherty Charitable Foundation 51
Herron, Ellen 109–111, 114
Hess, Henry 44–45, 91–97, 99, 101, 103, 114
Holmes, Arthur 115–116
hot spots 93, 97
hot springs 5
Hudson Submarine Canyon 53–54
Hutton, James 25–28

I

Iceland x, 16, 34, 40, 61
Icelandic ponies 34, 36
Imperial College of London 84
Indonesia 14
Inuits 34, 36, 40
iron 4, 21, 67–69, 84

J

Japan ix, 14
Jaramillo Event 76, 78, 111–*113*
Jardetzky, W. S. 62
Jeffreys, Sir Harold 115
Jensen, James 119
jet stream 40
Johnson, Lyndon 62
Journal of Geological Research 117
Juan de Fuca Ridge-Rift 86–*88*, 89, 99–100, 102, 109–110, *113*

K

Kanai Island 97
Kelvin, Lord (William Thomson) 28
Kilauea, Mt. 97
Koch, Johann Peter 34–36
Koppen, Else 36–37
Koppen, Vladimir 37

L

laboratories, scientific 16, 75
La Jolla, California 79
Lamont-Doherty Geological Observatory 51, 62 *See also* Lamont Geological Observatory
Lamont Geological Observatory 48, 56–57, 63, 70–71, 81, 83, 104, 109–113
    Ewing's leadership of 51–52
    Heezen and 59–61
    Opdyke and 106–107
Lamont, Thomas 51

land bridge 32
lava *See* magma
lead 70
Lehigh University 50–51
Leonardo da Vinci 25
limestone 2
lodestones 67–68
Loewe, Fritz 40–41
Luzon Island ix
Lyell, Charles 27–28
lystrosaurus 119

M

magma 4–9, 12, 14, 21, 81, 92–94, 111
magnetic field 4–5, 77, 87
magnetic poles 63, 72, 107
magnetic stripes 82, 87, 88–89, 99–100, 102, 113
magnetism 21–22, 24, 37, 52, 61, 63, 87, 102, 107, 113
    interpretation of 99–100
    in volcanic rock 67–71, 72–77, 78–79
    on the ocean floor 84–85
magnetometer 69, 73, 79, 82, 84–88, 107–109
mantle, Earth's 3–5, 11–12
Massachusetts Institute of Technology 64, 66
Marquess Fracture 83
Mason, Ronald 84–87, 89, 100, 108
Mathews, Drummond 97, 99–100, 109, 112–113
Maury Trough 83
McDougal, Ian 74–75
McKenzie, Dan 117
Menard, William 81–84, 87, 89–91, 93, 96, 101–103, 114, 116–117
Mendocino Fault 82–83, 89
Menlo Park, California 66–67, 74, 79, 111–112
meteorites 90
Mid-Atlantic Ridge 5–6, 8, 16, 53, 57–59, 83, 117
MIDPAC 80–81, 91
Midway Island 97–98
minerals 30, 115
moon 32, 90
Morgan, W. Jason 116–117
Morley, Lawrence 99–100, 109, 112–113
mountains, origin of 28–29
Murray Escarpment 83
Mylius-Erichsen, Ludvig 33, 39

N

National Academy of Science 62
National Aeronautics and Space Administration (NASA) xii, 112, 116
National Science Foundation xii
natural disasters ix–x
Nature 100, 117
Navy Electronics Laboratory 81, 91

Nazca Plate 10–11
Newfoundland 54
New Jersey 45
New Mexico 76
Noah, biblical story of 25
North American Plate 14
North Pole 37
nuclear science 23

O

ocean currents 52
ocean floor, mapping of 3, 19, 20, 50–53, 57–59, 81–85, 86–89
ocean floor, structure of xii, 9–11
oceanography 17, 44–46, 50–53
Office of Naval Research 80
Olduvai Event 76, 78, 111, 113
Olduvai Gorge 76
Opdyke, Neil 70, 72, 107–110, 112, 114
Origins of the Continents and Ocean, The (Wegener) 37–38

P

Pacific Plate 2, 11, 97–98
Parker, Robert 117
passive margin 13–14
petroleum 30, 55, 64, 115
Philippines ix
Pinatubo, Mt. ix
Pitman, Walter 109–112, 114
plate tectonics v–vi, 12, 21, 24, 59, 98
    confirmation of 117–120
    Ewing and 61–62
    theory of xi, 2–3, 5–8, 105–106
Playfair, John 27
plugs 17, 18–19, 52, 84–85, 106–109
plumes, magma 5–7, 12
polar reversal, magnetic 84, 87, 99, 102, 106, 112
    and plate tectonics 108–109
    in volcanic rocks 72–77, 78–79,
polar wandering, magnetic 70–71, 72–73
poles, geographic 68
poles, magnetic 68–73
potassium 70, 75
Press, Frank 62
Princeton University 44–45, 48, 50, 82, 91, 96, 101, 112, 116
Principles of Physical Geography (Holmes) 115
productive zones 10
protocracks 101

Q

quartz 53

R

radioactivity 23, 28, 69–70

Raff, Arthur 84–87, 89, 100, 108
red clay 84–85, 108
research methods 16–20, 23
Revelle, Roger 80–82, 85, 87
Rice Institute 50
Ridge Crest 7–8
ridge-rift 7–8, 10, 58–59, 83, 86, 92, 101–102, 109–110
rift 5–8, 14–15, 57–59, 101
Rock Magnetic Project 66
Royal Canadian Army Engineers 101
Royal Navy 103
Royal Prussian Aeronautical Observatory 33
Runcorn, Keith 70, 72, 107
Rutherford, Lord (Ernest) 103
Rutten, Martin 73

S

St. Helens, Mt. *xi*, 9
San Andreas Fault 29, 101–102
Scapegoat Mountain, Arizona 26
Science 116
science and society v, xii
scientific methods 50, 52
Scripps Institution of Oceanography 48, 79–82, 84–85, 91, 100, 103, 108, 117
seafloor spreading 14, 21, 82–83, 94, 100, 110, 114–115
  geometry of 105–106, 118
  theory of v, x, 96–97, 112
seamounts 45, *88*
sediment 2, 10, *18*, *26*, 50, 53, 81, 84
seismic profiler 57–59
sonar 52 *See also* echo sounding
sound waves *See* echo sounding
South American Plate 5, 10
South Atlantic Ridge-Rift 10
Space Science Board 96
striations 26
subduction 9–10, 107, 111
submarine warfare 46
subsidence 27, 29
supercontinent 12, 32
Sykes, Lynn 111–112, 114

T

Tharp, Marie 57–59
trench, oceanic 9–*11*, 93, 101–102
triple junction 14
turbidity currents 55

U

Unzen, Mt. ix

U.S. Army Air Corp 90
U.S. Coast and Geodetic Survey 46, 50, 82, 96
U.S. Congress xii
U.S. Geological Survey 48, 61, 65–66
U.S. Marine Corp 45
U.S. Navy v, 45, 51, 61
U.S. Office of Naval Research 46, 91
University of California at Berkeley 48, 63–67, 74
University of California at Los Angeles 63–64
University of California at San Diego 48
University of Chicago 38
University of Illinois 90
University of Marburg 34, 36
University of Oklahoma 64
University of Toronto 48, 64, 100–101, 103
uranium 69–70

V

Vacquier, Victor 87, 89
Vema (research vessel) 55–56, 59
Verhoogen, John 64, 66
Vine, Fred 97, 99–100, 109, 112–113
Vine-Mathews-Morley hypothesis 100
Virginia 50
Virgin Islands 44
volcanoes ix–*xi*, 1, *3*–5, 9, *11*–12, 14–*15*, 29, 58, 91–95, 97, 115

W

Warhaftig, Clyde 66
Wegener, Alfred 32–*41*, *42*–44, 49, 61, 90, 104–105, 114–115, 118–120
Wegener, Kurt 33
Wilkie, Charles 56
Wilson, James T. *See* Wilson, Tuzo
Wilson, J. T. 100
Wilson, Tuzo 100–102, 107, 111
women in geology 57
Woods Hole Oceanographic Research Center 45, 48, 51–53, 96
World War I 36–37
World War II v, xii, 44–47, 51, 64, 80, 90–91, 103

Z

zebra stripes *88*, *113* *See also* magnetic stripes